The literature on the evolution, taxonomy, anatomy, physiology and genetics of citrus is voluminous and spread between many different publications. This book aims to provide a concise, up-to-date, comprehensive and critical overview of the biology and cultivation of citrus, accessible to all concerned with citrus biology and culture.

The book begins by considering the origin and history of citrus from antiquity to modern times. The distribution of various citrus crops is then described, leading to a discussion of the taxonomy of citrus and the horticultural classification of the main citrus species. Particular attention is paid to problems of taxonomy within the genus *Citrus* and the contribution that the techniques of molecular biology have made towards their solution. A description of the vegetative and reproductive anatomy of citrus precedes a full discussion of reproductive physiology, dealing with flowering, fruiting, productivity, ripening, postharvest and fruit constituents. The main aspects of cultivated citrus, such as rootstocks, irrigation and mineral nutrition, pests, viruses and diseases are dealt with, leading to a concluding chapter that considers the potential for genetic improvement, including the use of tissue culture and plant biotechnology.

Biology of Citrus includes many original illustrations and offers lists of recommended reading as well as extensive references to the primary literature, making it ideal as an essential reference work for students and citrus specialists.

T0269322

BIOLOGY OF CITRUS

BIOLOGY OF HORTICULTURAL CROPS

Existing texts in horticultural science tend to cover a wide range of topics at a relatively superficial level, while more specific information on individual crop species is dispersed widely in the literature. To address this imbalance, the *Biology of Horticultural Crops* presents a series of concise texts, each devoted to discussing the biology of an important horticultural crop species in detail. Key topics such as evolution, morphology, anatomy, physiology and genetics are considered for each crop species, with the aim of increasing understanding and providing a sound scientific basis for improvements in commercial crop production. Volumes to be published in the series will cover the grapevine, citrus fruit, bananas, apples and pears, and stone fruit.

The original concept for this series was the idea of Professor Michael Mullins, who identified the topics to be covered and acted as General Editor from 1983 until his untimely death in 1990.

BIOLOGY OF CITRUS

Pinhas Spiegel-Roy

Department of Fruit Breeding and Genetics
Volcani Center, A.R.O. Bet Dagan, Israel

Eliezer E. Goldschmidt

The Kennedy-Leigh Center for
Horticultural Research
Faculty of Agriculture, Rehovot
The Hebrew University of Jerusalem, Israel

CAMBRIDGE
UNIVERSITY PRESS

CAMBRIDGE UNIVERSITY PRESS
Cambridge, New York, Melbourne, Madrid, Cape Town, Singapore, São Paulo

Cambridge University Press
The Edinburgh Building, Cambridge CB2 8RU, UK

Published in the United States of America by Cambridge University Press, New York

www.cambridge.org
Information on this title: www.cambridge.org/9780521333214

First published 1996
This digitally printed version 2008

A catalogue record for this publication is available from the British Library

ISBN 978-0-521-33321-4 hardback
ISBN 978-0-521-05424-9 paperback

Contents

Preface

The present book aims to provide a concise, up-to-date reference book on most aspects of citrus biology. Citrus is second only to the grape (the largest area of which is planted for wine) as a fruit crop and has been the subject of many studies. Six volumes of *Citrus Industry* issued by the University of California, dealing in detail with many aspects of citrus, have been published since 1967. Important information can be gathered also from the *Proceedings of International Congresses of Citriculture*. Our book provides an introduction to and overview of underlying principles and findings of citrus biology and culture. However, some important topics may have been omitted or treated too briefly. Emphasis has been placed on up-to-date treatment and conceptions of citrus physiology, reproductive development, taxonomy, genetics and breeding. The extensive references accompanying each chapter, including recommended reading, will be helpful to the reader, though they are of course, far from complete. Illustrations have been provided throughout to accompany the text.

Certain prominent up-to-date aspects of cultivated citrus are contained in a separate chapter devoted to the subject. The book also contains a contribution by Prof. D. Rosen, of the Faculty of Agriculture, the Hebrew University, on citrus pests, which has been specifically written for this book. The book will be useful to undergraduates, as well as to students in advanced courses, specializing in citriculture and horticulture. An understanding of elementary plant sciences is being assumed. We hope that this book will also appeal to citrus specialists as well as to general readers.

P. Spiegel-Roy
E. E. Goldschmidt

Acknowledgements

The completion of this book has been delayed by the death of the original co-author Prof. S. P. Monselise. Meanwhile the support of the editorial staff at Cambridge University Press has been unwavering. The authors thank Dr Aliza Vardi, Prof. Y. Shalhevet, Dr Y. Erner, Prof. D. Zohary, Prof. M. Bar-Yoseph, Dr Z. Solel, Dr D. Orion and Dr A. Shaked for reviewing and providing most helpful comments on various chapters of the book

Special thanks to Prof. M. L. Roose, University of California, Riverside, for valued assistance with illustrations for the chapter on Citrus and its relatives.

We would like to acknowledge the clerical skills and hard work of Mrs R. Gothard and Mrs N. Ben Yehezkel who were responsible for the typing of the manuscript, to Mrs L. Rosentul and Mrs N. Gestetner for illustrations (artwork). We would like also to thank Mr O. Tevel and Mrs Z. Sadovsky for photography of plant material.

Introduction

THE BIOLOGY, HISTORY and development of citrus fruits have aroused worldwide interest. This has been enhanced, to a large degree, by the uniquely attractive appearance of the fruit and by its medicinal properties.

Citrus fruits originated in South East Asia and spread during the Middle Ages, later to become established in all continents. Citrus is by far the most important evergreen fruit crop in world trade. The fruit's special structure and long shelf life have facilitated its large-scale export as fresh fruit. Processed juice products, on the other hand, have also become increasingly important worldwide.

The exact origin of *Citrus*, its ancestral types and systematics are still largely unknown. The great wealth of citrus types and cultivars of today reflects the vast natural breeding options within *Citrus*, as well as effective intentional human intervention. Molecular genetics, which has been a most helpful tool in unraveling the secrets of the past, also opens new vistas for breeding work in the future. Modern citriculture has adopted parthenocarpy and seedlessness for all major citrus types. Present-day cultivars represent largely subtle gene combinations conserved by vegetative propagation on seed-propagated, apomictic rootstocks.

Classical citriculture achieved the highest fruit quality in subtropical areas. Low temperatures and frost hazards limit the expansion of citriculture into cooler domains. Citrus has always been known to be highly dependent on irrigation in most environments. Water relations and mineral nutrition have been extensively investigated. The significance of viral diseases for the survival and propagation of citrus has become increasingly evident during the present century. The noticeable increase in marketing standards has dictated the adoption of strict disease- and pest-control strategies, involving extensive use of pesticides and fungicides. The development of biological pest-control measures is one of the most important achievements in this area.

The vegetative and reproductive physiology of citrus has been studied in great detail, with fruit growth and maturation receiving considerable attention. Prolonged storage and the overseas export trade have stimulated the study of postharvest physiology and pathology. The rise of the processing industry promoted detailed investigations into the chemical composition of the fruit.

Pressure for change

Citrus, in common with many other fruits, has expanded worldwide and is subject to universal trade. Citrus production is concentrated in many subtropical climates with irrigation (Mediterranean, California, South America, S. Africa, Australia) and is found to an even larger extent in the humid subtropical climates of Brazil and Florida. Particular developments in the cooler climate of Japan and expansion in the ancient habitat of China have also taken place. Marked differences have been observed in tree and fruit response under varying climatic conditions in arid, subtropical, tropical and marginal environments.

The search and adoption of seedless cultivars can be considered a major innovation. The effects of rootstock and of juvenile characteristics seem to be pronounced to a much larger degree than in any other fruit tree.

Significant progress has been achieved in irrigation and fertilization practices, pests and disease control, fruit storage, shipment and processing and cold protection.

Noteworthy progress has occurred in the field of citrus protoplast culture and fusion, and, recently, also in transformation techniques and in the contribution of molecular biology to the study of viruses.

Citrus culture has evolved in many areas with outstanding success; however, costly failures and weighty problems also abound.

Citrus groves are almost universally threatened by climatic hazards, viruses and decline diseases. Pressures for changes in cultivars, rootstocks and cultural practices are large. The biological control of insects, which has had a success unparalleled in other crops, faces new tasks. Soil and irrigation water salinity abound in many otherwise suitable environments. Further demands for changes in methods of production will prevail through economic and social forces. Chemical crop protection is both expensive and disquieting to the public. New pests and diseases emerge and are being disseminated. Though noteworthy progress has been achieved in the control of many destructive viruses, problems caused by certain viruses still greatly influence rootstock and scion selection. The replacement of chemical control by genetically engineered resistance emerges as a subject of high priority. Limitations on the manifold uses of growth substances in citriculture may well also be imposed in the future.

Social and economic pressures also call for the simplification of cultural methods, earlier fruit bearing and, possibly, new fruit types and products. While the selection and, recently, the breeding of seedless fruit has progressed measurably, aspects of self compatibility, pollination requirements and various aspects of seed formation still require further elucidation.

In contrast to the success achieved in affecting external fruit appearance and freedom from blemishes, successful influence on the internal quality and composition has been rather elusive.

The picking of citrus fruit involves substantial labor. Attempts to introduce large-scale mechanical picking or to facilitate fruit harvesting by chemical means (such as ethylene secreting substances) have not yet been successful enough. An increasing share of the world's citrus crop is consigned to industrial purposes, along with a steadily diminishing proportion for fresh fruit use, a trend having widespread implications as to choice of cultivars and management practices.

This book was written in a belief that a comprehensive compact, up-to-date exposé of citrus biology will contribute to the study, teaching and exploration of citrus.

I

History and growing of citrus

History of citrus

THE TERM CITRUS originated from the Latin form of '*Kedros*', a Greek word denoting trees like cedar, pine and cypress. As the smell of citrus leaves and fruit was reminiscent of that of cedar, the name citrus has been applied to the citron. Linnaeus grouped all citrus species known to him in the genus *Citrus*. In Greek mythology citrus fruits are called hesperides.

The suggested origin of the true citrus fruits is South East Asia, including South China, north-eastern India and Burma. Evidence from wild citrus in the area is still unclear. In many cases, seed has been spread large distances from the sites of origin and culture by birds, water streams and human activity. Tolkowsky (1938) considers the centre of origin to be the mountainous parts of southern China and north-eastern India, where sheltered valleys and southern slopes are protected from cold and dry winds yet are exposed to the warm rains of the summer monsoon. The deciduous *Poncirus trifoliata* grows wild in central and northern China.

While according to certain authors (Tanaka, 1954; Jackson, 1991) citrus fruits may have originated in north-eastern India and Burma, the introduction of citrus into cultivation and the probable origin of several species started in China. Table 2.2 gives the principal species of *Citrus* and their probable native habitat, according to Cooper and Chapot (1977). Lemon and grapefruit are not considered true species.

Domestication could have started independently in several locations in the area mentioned above or even in a broader area. There are indications of early cultivation of citron in India, and of mandarins and possibly other citrus fruit in China. The wide diversity of citrus in Yunnan has recently been described (Gmitter and Hu, 1990). Rivers arising in or traversing Yunnan could have served as dispersal mechanisms to the south.

Ancient dynasties of China regarded citrus as highly valued tributes

(Tolkowsky, 1938; Webber, 1967; Needham, 1986). The earliest mention of citrus fruits in Chinese literature occurs in the list of tribute articles sent to the imperial court at An-Yang (near the big bend in the Yellow River) during the reign of Ta Yu (*c.* 2205–2197 BC), as given in the *Shu Ching* under the section entitled *Yu Kung* of the Chinese Imperial Encyclopedia (Cooper and Chapot, 1977). The text may be as old as the early eighth century BC or the late ninth century BC (Needham, 1986). Figure 1.1 shows wrapped tributes of *Chu* and *Yu*, sent from Yanchow. The term *Chu* included presumably both kumquat and small-fruited mandarins. The term *Yu* included probably both the pummelo (*Citrus maxima*) and the Yuzu (*C. junos* Sieb. ex. Tan.). *Chu* and *Yu* tributes were perishable and may therefore have been essentially mandarins and pummelos. One of the earliest traditions of the Chinese was the grafting of *Chu* onto *Chih* (the deciduous *Poncirus trifoliata*). Pummelo was probably multiplied by air layering (marcottage). There was no mention of *kan* in the list of tributes. *Kan* may have included large mandarin-type fruit and possibly also oranges. The first mention of *kan* was made in a prose poem by Ssu Hsiang-ju, who died in 188 BC. From the time of WuTi (140–87 BC) Canton had an official in charge of the imperial tribute of yellow- and red-skinned *kan* orange-type fruit. Mention of the sour orange (described as unfit to be eaten in the raw state) and kumquat (*Fortunella*) also appears in the above-mentioned period (118 BC). The first description of citron appears only later, by Chi Han (AD 290–307). Citron fruits shaped like melons are described. Citrons originated most probably in India. Recently, a report of citron trees claimed to be indigenous has come from China. Han Yen Chih, in a monograph on citriculture in 1178 AD, described 27 cultivars of citrus. The earliest reference to citrus fruits in India appears in a collection of devotional texts around 800 BC, named *Vajasaneyi samhita*. Lemon and citron are specifically mentioned under the name of *jambila* (Tolkowsky, 1938). Names for oranges appeared in India for the first time in the oldest Sanskrit medical work about 100 AD (Tolkowsky, 1938). The Sanskrit name *nagarunga* has become *aurantium* in Latin, and orange in English (Jackson, 1991).

The role of the Shan race, also known as Tai, who were once dominant in Southern China and then forced to move west and southward into Burma and Assam, in producing and propagating and distributing citrus is not fully known. *Kan* oranges are still propagated today by seed in the Khasi hills of Assam.

The denotation of *Yau* in ancient Cantonese means pummelo fruit of the Yau people (the Tau race). High-quality pummelo cultivars and culture have developed in Thailand, where Tai people settled between the eighth and thirteenth centuries AD, essentially following the mode of propagation

Figure 1.1   Tribute of citrus fruit described in the *Yu Kung* chapter of *Shu Ching*. A late Ching representation from the *Shu Ching Thu Shuo*. From Needham (1986), reproduced with permission

(air layering) and culture of pummelo adapted in the flooded delta regions of China.

The sweet orange probably originated as a natural hybrid between the two species, pummelo and mandarin, grown in China in mixed village gardens. Human activity and interference of habitats confounded evidence that sweet oranges originated in the tropical rain forests of Upper Burma and Assam (Hooker, 1897; Tanaka, 1954). Probably, orange culture migrated from Yunnan to Upper Burma and eventually to Assam (Cooper, 1989). Sweet orange seed, which is apomictic, may have become naturalized later in the tropical rainforest region. It is postulated that sweet orange migrated from south-west China to Upper Burma rather than by the reverse path (Cooper, 1989).

The citron (*C. medica*), which is probably native to India, was not mentioned in Chinese writings until the fourth century AD. The citron was the first citrus with which the Europeans got acquainted and was perhaps for many years the only one known. The establishment of the citron in Media (Persia) appears to have occurred not later than the first half of the first millennium BC. It is assumed that it was first introduced by Alexander the Great to the Near East and Greece and it has been described by Theophrastus and called Persian (Median) apple. Evidence that citron was established at an earlier period in Egypt and Mesopotamia is rather inconclusive (Tolkowsky, 1938). Citron was appreciated for its medicinal properties, as an ornamental, for its fragrance and as an antidote to poison. It played a prominent part in Jewish religious rituals, appeared on Jewish coins during 66–70 AD, and has been a favorite motif in Jewish art since then (Figure 1.2). Caesarea was one of the main centres of citron culture. It was widely grown in Italy during the Roman period, probably as early as the first century AD.

A sculpture dating from the classic Hellenistic period clearly depicts lemon fruit, as well as the citron (Figure 1.3). The lemon, in addition to the citron, was known to the Romans, as evidenced by several mosaics and frescoes from the Roman era, including a mosaic from Tusculum dated 100 AD (Calabrese, 1990).

The Arabs were also well acquainted with the sour orange, and they have been instrumental in expanding citriculture to many areas. Expansion of sour orange culture occurred not later than the tenth century AD. Albertus Magnus (1193–1280) described the sour orange, calling it Arangus. By about 1150 AD citron, sour orange, lemon and pummelo had been introduced by the Arabs into Spain and northern Africa.

There is no written evidence of the actual culture of sweet orange in Europe before the fifteenth century AD. There are certain signs of earlier culture of the sweet orange (Tolkowsky, 1938). Citron, lemons and orange fruit, attached to cut branches, are clearly depicted in a mausoleum built by Constantine the Great (274–337 AD). Evidence of lemon and orange cultivation in Italy during Roman times is, however, still contested (Tolkowsky, 1938; Webber, 1967). Sweet orange presumably reached Europe through the commercial route established by the Genoese. The fruit was of the low-acid, sweet type (orange douce). The Portuguese contributed further to the spread and cultivation of the orange by introducing a superior variety with a more balanced flavour (Figure 1.4). The name Portugal clung to the sweet orange and was so adapted in various languages (*portogalea* by the Greeks; *burtugan* or *bortugan* in Arab countries). This may have stemmed from the belief that the sweet orange tree from the original introduction was still growing in Lisbon, in the garden of the Count St Laurent (Gallesio, 1811). The Portuguese introduction certainly had a profound influence on citrus industry and trade.

The lime is probably native to the East Indian Archipelago. The first mention of it in Europe was in the thirteenth century (Webber, 1967).

The mandarin is a native of China and may have been grown there for thousands of years. It was introduced to Europe fairly recently – in 1805 from China to Great Britain and from there to Malta. It has assumed increasing importance in the European market only during the second half of the twentieth century.

The mandarin is the foremost citrus in Japan. The first reference of

Figure 1.2 Citron fruit, mosaic floor of the Church of Nativity in Bethlehem (Fourth century AD). Israel Antiquity Authority

mandarin (Orange of Wenchou) in Japanese literature was made by Kokwan (1278–1346 AD). The famous Satsuma of Japan was named *Unshu-mikan* about 300 years ago. Tanaka (1932) suggested that it probably originated as a chance seedling in Japan during the Tang dynasty (618–907 AD). Mandarin seeds were brought to Japan from China, probably to Kagoshima on Kyushu island. A 300-year-old Satsuma tree

Figure 1.3 Cornucopia of the classical/Hellenistic period, in the National Archeological Museum of Athens, reproducing a citron (in the middle) and a lemon (on the right). After Calabrese (1990)

was found in Azuma-cho in Kagoshima, Kyushu. According to Chinese sources, a Buddhist priest brought back to Japan seeds from Unshu, China, giving rise to 'Satsuma' (M. Iwamasa, pers. commun.).

The kumquat (*Fortunella*) was introduced to the Royal Horticultural Society from China by Robert Fortune in 1846 AD.

In its journey to Europe the pummelo probably followed a similar path to that of the sweet orange and the sour orange. The Adam's apple, a form of shaddock, was mentioned as growing in the Holy Land around 1187 AD (Tolkowsky, 1938). It was brought to Spain by the Arabs at about the same period.

The pummelo is now widespread in Java, Malaysia, Thailand and Fiji. Some authors claim that it may have spread from the Malayan and Indian archipelagos to China and not vice versa. Calabrese (1994) states that the pummelo is of tropical origin (Malayan archipelago).

There are indications that seed of pummelo was brought to Barbados by Captain Shaddock, Commander of an East India ship. Pummelo tree and fruit is now often called 'Shaddock'.

Grapefruit (*C. paradisi*), now also classified as *C. maxima* var. *racemosa*, is almost certainly a hybrid of pummelo. It originated in Barbados, and was first described under the name of 'forbidden fruit' by Griffith Hughes from Barbados in 1750 (see also Gmitter, 1995).

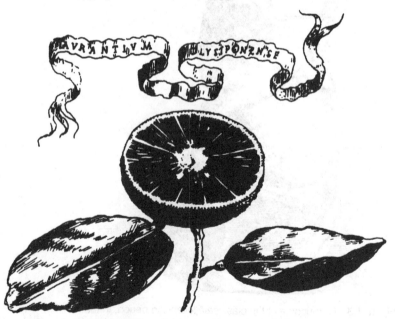

Figure 1.4 The Portugal orange as represented by Ferrarius in Hesperides (1646), named *Aurantium olysiponense*

Function and establishment of orangeries

Frost injury caused problems in establishing citrus trees in Europe. Citrus trees and fruits were highly regarded and prized. Hence, considerable efforts and expense were spent to enable their successful growth under adverse environmental conditions. As early as the first century BC, Seneca mentions use of panes of mica for protecting delicate plants in Rome. Special houses, known as *stanzone per i cidri* and later as orangeries, were established in the fourteenth century for culturing citrons and oranges in many parts of Europe (Figure 1.5). The orangeries can be considered as predecessors of the greenhouses and of greenhouse culture of various horticultural plants and produce.

Introduction of citrus into the Americas, Africa and Australia

No *Citrus* species is indigenous to America. The grapefruit, at present assumed to be a hybrid of pummelo with the orange, originated in Barbados. Travellers and missionaries greatly assisted the spread of citrus. As it became known that fresh fruits provide the best remedy against scurvy, the British Navy made it compulsory for sailors to drink lime juice daily. It has been documented that Columbus took from the Canary Islands seeds of oranges, lemons and citrons on his second voyage (1483), which landed at Hispaniola (Haiti). The first mention of citrus in the continent of America is in a manuscript written in 1568 and discovered in the archives of Guatemala. Seeds were brought from Cuba to Vera Cruz in 1518. Spanish and Portuguese ships helped in the establishment of citrus seed and subsequently trees along the voyage. Citrus was also successfully established in the West Indies and Brazil by the middle of the sixteenth century. It has since become abundant and even feral in some localities. Citrus was first brought to Florida some time between 1513 and 1565, with the first written reference on oranges dating from 1579. Wild-growing citrus trees and groves became established on hammock lands near lakes or rivers from seed dropped by native Americans. The first grapefruit seeds were probably brought to Florida not later than 1809 (Hume, 1926) or 1823 (Robinson, 1933) by Counte Odette Phillippi.

Citrus fruits reached Arizona before settlements were established in California (around 1707). They were introduced to California probably by 1769, when the first mission of the Franciscans was established in San Diego. The principal orange cultivar in the world, Washington navel, was introduced to Riverside, California in 1873.

Portuguese discoverers of the all-sea route to India, in around AD 1500, found oranges, lemons and citrons cultivated in several places in East Africa. Establishment of citrus may have been initiated some centuries before then. In east Africa, Arab and Indian merchants were instrumental

Figure 1.5   An orange house. From Ferrarius, Hesperides (1646)

in establishing citrus. In west Africa, including regions of Congo and St Helena, citrus was introduced by the Portuguese. The first sweet orange trees reached South Africa from St Helena in 1654, and were planted in the Dutch Governor's garden.

Citrus was first planted in Australia by colonists of the First Fleet, who introduced seeds and plants from Brazil in 1788 to New South Wales. Mandarins introduced from China were already growing in New South Wales by 1828.

Citrus production, by-products and trade

Citrus is second only to the grape (of which most is used for wine) in the area planted and in the production of fruit trees. Citrus plantings (FAO Statistics) amount worldwide to over two million hectares with citrus production estimated in 1992/3 at 76075000 tons (Table 1.1). Brazil is by far the largest producers of oranges (19.7%), followed by the USA (13.4%), China, Spain, Mexico, Italy, India and Egypt (Table 1.1). The largest producers of mandarin are Japan (mostly 'Satsuma') and Spain (mostly 'Clementine'), followed by Brazil, Korea, Italy, Turkey, USA, Pakistan, India, China, Morocco, Egypt and Argentina. A sizeable quantity is consumed in tropical regions and does not figure in the statistics. The largest producer of grapefruit is the USA, followed by Israel, Cuba, Argentina, South Africa and Cyprus. The largest producers of lemon are the USA, Italy and Spain, followed by Turkey and Greece. Mexico is the world's largest producer of lime.

Recent trends of citrus production and demand include all-year-round supply, the increasing importance of industrial products (mainly concentrated fruit juice), demand for seedless fresh fruit with a substantial increase in easy peeling, mandarin-type fruit and a growing demand for pigmented grapefruit (for fresh fruit). Navel orange is now available during the whole year, through supply from both hemispheres and the contribution of various mutants with different ripening periods.

Citrus has many uses, besides fresh fruit and consumer-processed fresh juice. Some of the uses are by-products of the processing industry and its main product – concentrated fruit juice. Products include canned fruit segments (mainly grapefruit and satsuma segments), citrus-based drinks, pectin, citric acid, seed oil, peel oil, essential and distilled oil, citrus alcohol, citrus wines and brandies, citrus jams, jellies, marmalades and gel products. Citric acid is recovered mainly from lemons and to a lesser extent from limes and bergamot. Citrus by-product wastes are used as

Table 1.1 *World citrus production in 1992/93 (thousands of tons)*

	Oranges	Tangerines (mandarins)	Lemon and lime	Grapefruit and pummelo
World	54593	9378	7127	4979
Northern Hemisphere	34756	8199	5241	4490
United States	9249	352	834	2532
Mediterranean Region	10849	3566	2539	703
Greece	1061	78	169	—[1]
Italy	2218	515	785	6
Spain	3002	1521	737	30
Israel	377	116	18	383
Algeria	140	111	10	3
Morocco	831	316	9	5
Tunisia	106	51	17	9
Cyprus	128	Not given	42	120
Egypt	1261	341	309	—
Lebanon	263	Not given	—	—
Turkey	775	345	270	59
Former USSR	280	Not given	—	—
Japan	194	2019	—	—
Cuba	428	15	27	307
Mexico	2530	185	845	118
China	4834	771	153	340
India[2]	1840	Not given	560	50
Pakistan[2]	1100	Not given	—	3
Philippines[2]	9	46	49	40
Thailand[2]	56	Not given	—	225
Southern Hemisphere	19837	1178	1886	489
Argentina	660	338	605	181
Brazil	14974	553	741	25
Uruguay	135	53	46	—
Venezuela	451	—	—	—
Chile	Not given	—	95	—
United States[3]	822	—	—	—
Australia	572	Not given	35	31
South Africa	730	Not given	51	106

[1] —, production none or very limited.
[2] Data for 1991/2.
[3] California Valencia orange production is included in the Southern Hemisphere total.

From FAO (Food and Agriculture Organization of the United Nations), *Citrus Fruit Annual Statistics 1994.*

molasses for animal feed. Several flavonoid compounds are used by the food and pharmaceutical industries.

The utilization of citrus as ornamental plants often preceded its cultivation for fruit production (Continella *et al.*, 1994). The value of citrus as an ornamental is enhanced by the multitude of species, their special shape and features, their attractive foliage, their fragrant flowers and long lasting fruits.

Climatic limits to citriculture: the world citrus belt

The evergreen *Citrus* species and cultivars – oranges, mandarins, grape-fruits, pummeloes, lemons, limes and citrons – grow and produce fruit under rather varied climatic conditions, ranging in latitude from over 40° north (Corsica, Japan) to almost 40° south (New Zealand); from equa-torial, hot-humid climates through warm–subtropical and even cooler maritime climates. The citrus belt of the world is shown in Figure 1.6.

The sensitivity of the tree and fruit to frost, varying somewhat between species and rootstocks, is a major factor limiting the regions and localities where *Citrus* can be successfully grown. A sufficiently long, warm summer is also required to enable the fruit to grow and reach maturity. This constraint becomes important at the cooler margins of the *Citrus* growing-area (except for lemons, which can be consumed before full maturity). In the Mediterranean and similar climates with long dry summer periods, irrigation is required to maintain satisfactory tree growth and fruit development.

Although *Citrus* grows well in the tropics, most of the commercial

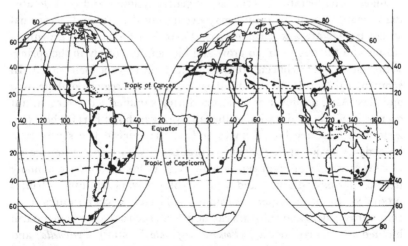

Figure 1.6 The world citrus belt. Commercially important regions have been shaded

citriculture is concentrated between the latitudes of 20° and 40°. The main difficulties of citriculture in the tropics are the distortion of the productivity cycle and the reduced fruit quality. In tropical equatorial regions, with high temperatures and humidity prevailing through the whole year, trees often tend to flower sparsely, resulting in lower productivity. Where periods of drought occur, trees burst into bloom following the rains that terminate the drought; in many cases this results in several crops during the year, a situation that is difficult to handle by the export and processing industries. While the uninterrupted high temperatures of the tropics enhance fruit growth and maturation, several aspects of fruit quality may suffer. The internal quality of fresh orange and mandarin cultivars may become inferior because of low acid content. Low temperatures are required for the development of the highly pigmented rind of oranges and mandarins, and their fruit in the tropics generally does not attain the desired colour. These factors are of less significance in yellow and high-acid cultivars (e.g. limes), which are indeed widely grown in the tropics. In addition, under the warm and humid conditions of the tropics fruits often suffer from rind blemishes and pests disfiguring their appearance. These factors have limited the expansion of commercial citriculture in the tropics. During the last few decades, however, due to improvements in cultural conditions and to the investment that has been made, the growing and marketing of citrus in many tropical areas has been considerably expanded. In the belt of subtropical latitudes (roughly 20° to 40° latitude north and south of the equator), which has definite seasons, the rhythm of blossoming and growth is controlled by seasonal changes in temperature. Humidity is generally lower and considerable daily changes in temperature may occur. Local frosts and occasional severe freezes are also encountered. There is a cessation of growth in winter and the trees start to grow and blossom uniformly in the spring, with a subsequent concentrated crop ripening (except for lemons). In semitropical Florida, Brazil and the citrus-growing regions of Argentina and East Asia, a rainy season occurs during summer; a dry season, if it occurs, is of shorter duration in the fall, spring or both. Part of the citrus zone is in frost-free humid zones and part is subjected to severe frost hazards. A major spring bloom is produced, with some out-of-season blooms caused by drought and rainfall. High yields are obtained in many parts of Florida and Brazil, and with suitable cultivars high-quality fruit is produced. In citrus regions constantly subjected to cold winters, only cold-hardy genera and cultivars can be grown, such as Satsuma mandarin budded to trifoliate orange (*Poncirus trifoliata* L. Raf.), *Fortunella*, and Bouquetier sour orange for processing for neroli oil.

Citrus cultivars with anthocyanin-colored rind and juice are successfully grown mainly in areas with low midwinter temperatures, as in Italy. Early-ripening cultivars attain the natural orange color during cool fall weather. Colored grapefruit, the color being mostly due to lycopenes, can be grown in a variety of climates. While anthocyanins develop in blood oranges during low winter temperatures, lycopene production in grapefruit is achieved mainly with prolonged fairly high temperatures. Thus, pigmented grapefruit attains an excellent color in Texas and Florida. For further discussion see also Chapters 2 and 4.

Recommended reading

Cooper, W. C. and Chapot, H. (1977). Fruit production – with special emphasis on fruit for processing. In *Citrus Science and Technology*, ed. S. Nagy, P. E. Shaw and M. K. Veldhuis, Vol. 2, pp. 1–127. Westport, CT: The Avi Publishing Co.

Gallesio, G. (1811). *Traité du Citrus*. Paris: Louis Fantin. 381 pp.

Hume, H. H. (1941). *The Cultivation of Citrus Fruits*. New York: Macmillan. 561 pp.

Tolkowsky, S. (1938). *Hesperides. A History of the Culture and Use of Citrus Fruits*. London: John Bale, Sons and Curnow. 371 pp.

Webber, H. J. (1967). History and development of the citrus industry. In *The Citrus Industry*, Vol. 1 ed. I. W. Reuther, H. J. Webber and L. D. Batchelor, pp. 1–39. Berkeley: Division of Agricultural Sciences, University of California.

Literature cited

Calabrese, F. (1990). The fabulous story of citrus fruit. *Agricoltura*, **208**: 82–128. (In Italian.)

Calabrese, F. (1994). The history of Citrus in the Mediterranean Countries and Europe. In *Proc. Int. Soc. Citriculture 1992*, ed. E. Tribulato, A. Gentile & G. Reforgiato, pp. 35–8. Catania, Italy: MCS Congress.

Continella, G., La Malla, G. and Romano, D. (1994). The utilization of citrus as ornamental plants in Italy. In *Proc. Int. Soc. Citriculture 1992*, ed. E. Tribulato, A. Gentile & G. Reforgiato, pp. 232–4. Catania, Italy: MCS Congress.

Cooper, W. C. (1989). *Odyssey of the Orange in China. Natural History of the Citrus Fruits in China*. Published by the author, 443 Lakewood Drive, Winter Park, Florida. 122 pp.

Cooper, W. C. and Chapot, H. (1977). Fruit production – with special emphasis on fruit for processing. In *Citrus Science and Technology*, ed. S. Nagy, P. E. Shaw and M. K. Veldhuis, Vol. 2, pp. 1–127. Westport CT: The Avi Publishing Co.

FAO Commodities and Trade Division (1993). *Citrus Fruit, Fresh and*

Processed. Annual Statistics. CCP; CI/ST/93. Rome, Italy: Food and Agriculture Organization of the United Nations.

Ferrarius, Giovanni Battista (1646). Hesperides; *sive de malorum aureorum cultura et usu libri quatuor.* Rome: Herman Scheus. 480 pp.

Gallesio, G. (1811). *Traité du Citrus.* Paris: Louis Fantin. 381 pp.

Gmitter, F. G. Jr (1995). Origin, evolution and breeding of the grapefruit. In *Plant Breeding Reviews,* Vol. 13, ed. J. Janick, pp. 345–63. New York: John Wiley & Sons.

Gmitter, F. G. Jr and Hu, X. (1990). The possible role of Yunnan, China, in the origin of contemporary *Citrus* species (*Rutaceae*). *Econ. Bot.,* **4**: 267–77.

Hooker, J. D. (1897). *The Flora of British India. Rutaceae,* Vol. 1, pp. 484–517. London: Reeve and Co.

Hume, H. H. (1926). *The Cultivation of Citrus Fruits.* New York: The Macmillan Co. 561 pp.

Jackson, L. K. (1991). *Citrus Growing in Florida.* 3rd edn. Gainesville: University of Florida Press. 293 pp.

Needham, J. (1986). *Science and Civilisation in China. Vol. VI. Biology and Biological Technology. Part I. Botany.* Cambridge: Cambridge University Press.

Robinson, T. R. (1933). The origin of the Marsh seedless grapefruit. *J. Hered.,* **24**: 437–9.

Tanaka, T. (1932). A monograph of the satsuma orange, with special reference to the occurrence of new varieties through bud variation. *Mem. Fac. Sci. Agr. Taihoku Univ.,* **4**: 1–626.

Tanaka, T. (1954). Species problem in *Citrus* (Revisio aurantiacearum, IX). *Jap. Soc. Prom. Sci.,* Ueno, Tokyo. 152 pp.

Tolkowsky, S. (1938). *Hesperides. A History of the Culture and Use of Citrus Fruits.* London: John Bale, Sons and Curnow. 371 pp.

Webber, H. J. (1967). History and development of the citrus industry. In *The Citrus Industry,* Vol. 1, ed. I. W. Reuther, H. J. Webber and L. D. Batchelor, pp. 1–39. Berkeley: Division of Agricultural Sciences, University of California.

2

Citrus and its relatives

THE TRUE CITRUS fruit trees belong to the family of Rutaceae, subfamily Aurantioideae. Rutaceae is one of the four families in Rutales, division Lignosae of the subphylum Dicotyledoneae, with mostly subtropical or tropical genera. Leaves usually possess transparent oil glands and flowers contain an annular disc. Rutaceae contains about 150 genera and 1600 species (Swingle and Reece, 1967).

Aurantioideae, the 'Orange' subfamily – one of seven subfamilies in Rutaceae (Engler, 1931) – has been subdivided by W. T. Swingle into two tribes – Clauseneae and Citreae (a single tribe according to Engler, 1896), with 33 genera and 203 species. The main characteristics of the subfamily are: the fruit is a berry (hesperidium) with a leathery rind or hard shell, often with juicy pulp in the subtribe Citrineae. The seeds are without endosperm, sometimes with two or more nucellar (apomictic) embryos. The leaves and bark have schizolysigenous oil glands. They are small or sometimes large trees, rarely shrubs. Incorporating new taxonomic information, the number of species is now estimated at about 220, though a reduction in according species rank to several citrus species will reduce this number (see Table 2.1). Relationships between genera have in the past been based mainly on comparative morphology. Morphological affinities have been supported to some extent by grafting and hybridization performance.

Citrus relatives

An increase in interest in the Aurantioideae has been shown in the quest for wild members of the subfamily as a source of novel genetic variation and as a possible source of rootstocks, as well as for traits for pest and disease resistance. A recent review on progress made in the taxonomic research on the Aurantioideae of South East Asia (where about two thirds of the species occur naturally) has been presented by Jones (1989). All the

Table 2.1 *Tribes, subtribes and genera of the subfamily*
Aurantioideae

Tribe	Subtribe	Genus	Species
Clauseneae	Micromelinae	*Micromelum*[1]	9
	Clauseninae	*Glycosmis*[1]	35
		Clausena[1]	23
		Murraya[1]	11
	Merrillinae	*Merrillia*[1]	1
Citreae	Triphasiinae	*Wenzelia*[1]	9
		Monathocitrus[1]	1
		Oxanthera	4
		Merope[1]	1
		Triphasia[1]	3
		Pamburus	1
		Luvunga[1]	12
		Paramignya[1]	15
	Citrinae	*Severinia*[1]	6
		Pleiospermium[1]	5
		Burkillanthus[1]	1
		Limnocitrus[1]	1
		Hesperethusa[1]	1
		Citropsis	11
		Atalantia[1]	11
		Fortunella[1]	4
		Eremocitrus	1
		Poncirus	1
		Clymenia[1]	1
		Microcitrus[1]	6
		Citrus[1]	16
	Balsamocitrinae	*Swinglea*[1]	1
		Aegle	1
		Afraegle	4
		Aeglopsis	2
		Balsamocitrus	1
		Feronia	1
		Feroniella[1]	3
2 Tribes	6 Subtribes	33 Genera	203

[1] Genera occurring naturally in South East Asia
(according to Jones, 1989).

After Swingle and Reece (1967).

species of the Aurantioideae are trees or shrubs with persistent (ever-green) leaves, except in three monotypic genera (*Poncirus, Aegle, Feronia*) and in three species of *Clausena* and one of *Murraya*. Flowers are usually white and often very fragrant. Many genera bear fruit with a green, yellow or orange peel dotted with numerous oil glands. The genus *Citrus* and few related genera have fruits with juicy pulp vesicles. The subtribe Bal-samocitrinae have fruits with a size similar to that of oranges but with hard woody shells.

Many remote relatives of *Citrus* bear extremely small fruits compared with citrus fruit. A few of them have been found to be graft-compatible with *Citrus*. Of the 33 genera belonging to the Aurantioideae, 29 are considered native to the Monsoon region extending from West Pakistan to north-central China and from there south through the East Indian Archipelago to New Guinea and Bismarck Archipelago, Australia, New Caledonia, Melanesia and the western Polynesian islands. Five genera are native to tropical Africa. One genus, *Clausena*, is native to both the Monsoon region and to tropical Africa. Some of the genera related to *Citrus* are of importance as ornamentals.

The 33 genera of the subfamily Aurantioideae are divided into two tribes: the Clauseneae – five genera, including what are considered remote relatives of *Citrus* – and the Citreae – 28 genera, incluing *Citrus* and closer relatives.

Clauseneae comprise the more primitive genera of the orange sub-family. None of the species develop spines in the axils of the leaves. The odd-pinnate leaves are easily distinguished from those of Citreae by leaflets attached alternately to the rachis. Fruits are usually small, semidry or juicy berries, except in *Merrillia*. In the latter, the fruit is of ovoid shape with a thick, leathery exocarp. Its flowers are the largest in the Aurantioideae (55–60 mm diameter); they are trumpet shaped, becoming pendant during anthesis. *Murraya paniculata*, with fragrant flowers and small red fruits, is grown as an ornamental in Asia and in greenhouses. *Clausena lansium* (the Chinese wampee) is cultivated for its edible fruit in southern China (Swingle and Reece, 1967).

In the tribe Citreae, nearly all species develop single or paired spines in the axils of the leaves of vigorous shoots. The leaves are simple, unifolio-late or trifoliate, but a few genera have odd-pinnate leaves with the leaflets attached to the rachis.

The tribe Citreae has been classified into three subtribes.

1 Triphasiinae – minor citroid fruit trees.
2 Citrinae – citrus fruit trees.
3 Balsamocitrinae – hard-shelled citroid fruit trees.

It has been claimed that remote ancestors of cultivated citrus trees may have been very similar to some of the Triphasiinae.

The subtribe Citrinae, with 13 genera, differs from other members of Aurantioideae by having pulp vesicles, structures arising from the dorsal wall of the locule, growing into the locular cavity and developing into sacs filled with large, thin-walled cells with watery juice. No such structures have been found in other plants of Rutaceae or related families. No close homologies are known in any of the higher plants. Other genera of Aurantioideae have secretory glands on locule walls, giving rise to mucilaginous gum filling the locular cavity of the fruit. Lack of pulp vesicles characterizes the subtribe Triphasiinae.

The 13 genera of the subtribe Citrinae (Table 2.1) have been classified into three groups (Swingle and Reece, 1967). Group A, also called 'primitive citrus fruit trees', comprises five genera. They possess primitive forms of pulp vesicles, which is of great interest in the study of their origin and evolution. *Burkillanthus* has an ovary with 22–26 ovules in each of the five locules. *Severinia*, with six species having stalkless, peripheral pulp vesicles, has been studied more extensively as a possible source for rootstock. Citrus plants grafted on *Severinia buxifolia* (Figure 2.1) and *Severinia disticha* have shown remarkable tolerance to excess boron in sand cultures. Some *Citrus* plants grafted on *Severinia buxifolia* have survived for 30 years. This monoembryonic species has also shown cold tolerance, resistance to the citrus nematode *Tylenchulus semipenetrans* and to *Phytophthora* rot. It is, however, intolerant to the tristeza virus. The possibility of raising *Citrus* on a plant taxonomically very remote has stimulated interest in testing further relatives of *Citrus* for rootstocks and in hybridization.

Citrus has been also grafted on *Hesperethusa crenulata* (Bitters *et al.*, 1969). Recently, somatic hybrids have been developed by protoplast fusion of *citrus* and several distant genera (see Chapter 6, on genetic improvement in citrus; Table 6.3).

Group B, 'near Citrus fruit trees', includes two genera only, *Citropsis* and *Atalantia*, which show well-developed pulp vesicles with broad sessile bases and conical sides tapering to the acute apex. Pulp vesicles are arranged radially with the bases at the periphery of the locules attached to the dorsal wall of the locules and inbedded in the inner layer of the rind. The conical pulp vesicles point toward the center of the fruit (unless deflected by the seeds). *Citropsis* is native to Africa, with 10 out of 11 species having pinnate or trifoliolate leaves (Figure 2.2). *Atalantia*, also with 11 species, is native to South East Asia. Leaves are unifoliolate or simple, resembling those of *Citrus*. *Atalantia* is a potential source of resistance to the citrus nematode (*Tylenchulus semipenetrans*) and *Citropsis* to the burrowing nematode (*Radopholus similis*). Graft compatibility with *Citrus* has been shown by several *Citropsis* species. Belgian horticulturists experimented in Congo with *Citropsis gilletiana* (Gillet's cherry orange), the largest species in *Citropsis*, as a rootstock. It has shown remarkable tolerance not only to the brown-rot fungus (*Phytophthora citrophthora*) but also to the larvae of a longicorn beetle, *Monohammus* sp., which aggravates the severity of foot rot caused by *Phytophthora*.

Typical species of *Atalantia* have sessile, broad-based, conical pulp

Figure 2.1   *Severinia buxifolia* (courtesy Dr Roose)

Figure 2.2 *Citropsis*. Scale is in centimeters

vesicles growing out from the dorsal locule walls, filling the locules. *Atalantia ceylanica*, however, shows very few pulp vesicles, as the very large seeds almost completely fill the locules. Two other species classified as *Atalantia*, *A. hainanensis* and *A. guillaumini*, are devoid of pulp vesicles altogether.

True citrus fruit trees

This group (C) includes six genera: *Fortunella*, *Eremocitrus*, *Poncirus*, *Clymenia*, *Microcitrus* and *Citrus*. All have orange- or lemon-like fruits with specialized slender, stalked, more-or-less fusiform pulp vesicles. The latter fill all the space in the segments of the fruit not occupied by seeds. The number of stamens is at least four times higher than that of petals. All genera except *Poncirus* (with trifoliate, deciduous leaves) have unifoliate or simple leaves. *Clymenia* has simple leaves borne on wingless petioles with prominent venation on the lower surface. *Clymenia* (Figure 2.3) differs also in having the majority of pulp vesicles attached to the dorsal wall of the locules. Many of the pulp vesicles are attached to the radial wall of the segments. All other genera in this group have fusiform pulp vesicles on extremely slender stalks.

The pyriform pulp vesicles of *Clymenia*, being neither slender and stalked at the base nor with acute apices, point to a status intermediate between Group B (near citrus fruit trees) and Group C (true citrus fruit trees). Moreover, pulp vesicles of all genera contain droplets of oil, with similar oil droplets evident also in *Citropsis* and *Atalantia*. Five of the six genera (except the little-studied *Clymenia*) have been reported to show successful grafting onto one-another, as well as hybridizing between genera. This is, however, far from being general. The situation has been reviewed by Barrett (1978). He has also reported that in crosses between *Microcitrus* and *Poncirus*, hybrid seeds do not germinate; *Fortunella* × *Poncirus* hybrids have been produced but they did not survive.

Figure 2.3 *Clymenia polyandra* (courtesy Dr Roose)

The region accupied by the six genera extends from north-eastern India and north-central China to east-central Australia and New Caledonia, and from Java to the Philippines. Some are also found in southern Japan. *Citrus* is native to the whole area, except in north-eastern Australia where *Microcitrus* and *Eremocitrus* are native. In northern China *Poncirus* is native. *Fortunella* occurs in south-eastern China along with *Citrus*; *Clymenia* occurs in the Bismarck Archipelago along with some species of *Citrus*.

Clymenia (Figure 2.3) is considered the most primitive genus in the group. It differs from *Citrus* in many important taxonomical characters. It has a type of pulp vesicle not found in any other citrus fruit. The leaves are also unlike those of other genera in the true citrus fruits, and the enlarged disk bears 10–20 times as many stamens as petals. The ripe yellow fruit is edible and it has even been mistaken for a sweet lime.

While *Fortunella*, and especially its subgenus *Protocitrus*, have been considered as the most simple and primitive end of a side branch ending in *Citrus* (Swingle and Reece, 1967), its classification as an independent genus remains somewhat questionable. The mitochondrial genome of *Fortunella* has been found to be indistinguishable from that of *Citrus* (Yamamoto et al., 1993). *Fortunella* differs from *Citrus* mainly in having two collateral ovules near the top of each locule (*Citrus* has 4–12). Though definitely evergreen, it possesses a degree of winter dormancy, enabling the tree to remain quiescent during weeks of warm weather without initiating growth or starting to flower. Of genera other than *Citrus*, only *Fortunella* has achieved commercial significance. This is because of its most attractive and edible fruit, known as the kumquat. *Fortunella margarita* and *Fortunella japonica* are quite widely cultivated in China, Japan and some subtropical environments. Fruits have a relatively thick, fleshy sweet and edible peel, and 4–7 segments filled with pleasant, mildly acid pulp. *Fortunella polyandra*, which is native to tropical regions, and is cultivated in the Malay Peninsula, has large globose fruits with a thin peel. *Fortunella hindsii*, which has very small globose fruits, is still reported wild in the mountains of Southern China. A hybrid of *Fortunella*, named calamondin, is of considerable importance as an ornamental plant, bearing showy citrus fruits (see Figure 2.4). Though accorded species rank as *C. madurensis* Lour, and *C. mitis* Blanco, it is probably a hybrid between a sour loose-skinned mandarin and the kumquat. Fruit of calamondin is widely used in the Philippines as a condiment.

While *Eremocitrus* (Figure 2.5) has ovary and fruit characters somewhat similar to *Fortunella*, it also has striking xerophytic adaptations, evident in the character of its gray-green, small leaves, with a thick cuticle and deeply sunken stomata. During a severe drought, leaves drop. The plants have very stout spines. Flowers are smaller, though similar to those of

Microcitrus. The ovary has three to five locules, with two ovules in each locule, as in *Fortunella*. Pulp vesicles in the fruit are less coherent than in the common citrus fruits. Seeds are monoembryonic. It grows wild in New South Wales and south-eastern Queensland, showing adaptation to cold, drought, salt and excess boron. *Eremocitrus glauca* has been successfully grafted with *Citrus* and the reciprocal graft is also possible.

Figure 2.4 Calamondin, *C. madurensis*

Microcitrus (Figure 2.6) is also semi-xerophytic and there are indications of it showing high stress tolerance. *Microcitrus* differs from *Citrus* by its dimophic foliage, free stamens, ovary with four to six locules, and coriaceous strongly veined leaves. *Eremocitrus* differs from *Microcitrus* by its thick leaves and cuticle, stomata on both faces of the leaves, an ovary with

Figure 2.5 *Eremocitrus glauca* (courtesy Dr Roose)

three to five locules, and two ovules in each locule. It is possible that the ancestral type from which both *Microcitrus* and *Eremocitrus* evolved was similar to *Microcitrus warburgiana*, a species found in New Guinea. Pulp masses in *Microcitrus* are filled with acid pulp and acrid oil. Evolution of

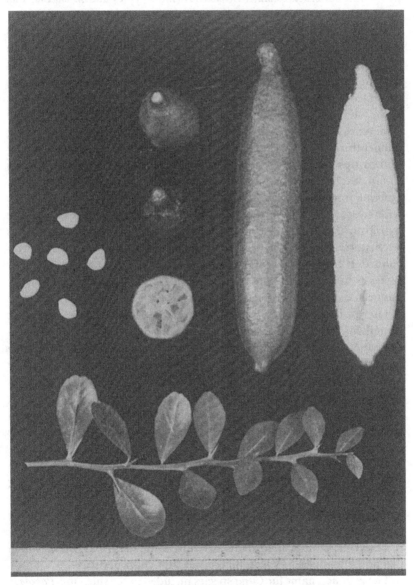

Figure 2.6   *Microcitrus australasica* var. *sanguinea* (courtesy Dr Roose)

Microcitrus occurred in New Guinea and Australia, which have been cut off from other land masses for 20–30 millions of years, and it is easier to follow than that of *Citrus*, *Fortunella* and Poncirus, which did not originate in regions that were geographically isolated during definitely dated geological periods. *Microcitrus* seems resistant to *Phytophthora* and the burrowing nematode, *Radopholus similis*. *Eremocitrus* (monotypic) and *Microcitrus* (six species) have been recently reviewed by Sykes (1993).

Poncirus is highly cold tolerant; it has been reported to withstand successfully even winter temperatures of $-26\,°C$. It has penetrated far into the temperate zone in north-eastern Asia. Leaves are trifoliolate and deciduous (Figure 2.7). Winter buds are well protected by bud scales. The protected flower buds form during early summer and bloom on old twigs in the following spring. Flowering is earlier than, at the same time as, and sometimes later than *Citrus* cultivars, depending upon winter and spring temperatures. Oil-containing pulp vesicles carry hair-like organs bearing at their tips thick-walled, fissured cells. Viscous fluid allows pulp vesicles to slip past one another. Immature fruit contains a glucoside, ponciridin, differing from hesperidin. *Poncirus* hybridizes readily with *Citrus*. Hybrids with sweet orange have been hybridized with *Eremocitrus*. Many of the hybrids with genera other than *Citrus* are sterile. *Poncirus* is widely used as a rootstock; in fact it may well be the most ancient rootstock used in fruit culture. It has been grown in China for thousands of years. In Japan, it serves as the main rootstock. A rather dwarf form, named 'flying dragon', has been recently experimented with as a rootstock, and has proven to be of interest in producing dwarf citrus trees. Hybrids of *Poncirus* are most prominent among new rootstocks bred for *Citrus*, carrying genes of resistance (from *Poncirus*) to tristeza virus, *Phytophthora*, citrus nematode and cold; in addition, as *Poncirus* is polyembryonic with predominantly apomictic (nucellar) offspring, it is suitable for propagation by seed. Affinity problems in grafting with *Citrus* have been noted with *Fortunella*, *Eremocitrus* and *Microcitrus*, but generally not with *Poncirus*. It is also grown as an ornamental in Asia and elsewhere, mainly in regions too cold for outdoor growing of citrus.

All four genera (*Poncirus*, *Eremocitrus*, *Fortunella* and *Microcitrus*) seem thus to possess valuable, though different adaptations to demanding climatic and soil conditions. They are therefore of increasing interest, mainly for the breeding of new types of rootstock, and possibly also for novel fruit types (Sykes, 1993).

The genus *Citrus* is divided into two subgenera, *Citrus* and *Papeda*, which can be distinguished by leaf, flower and fruit characteristics.

The common name for *Citrus* species included in the subgenus *Papeda* is Papedas. None of the species belonging to *Papeda* have edible fruits, as the pulp vesicles have dense aggregations of droplets of acrid oil. This subgenus has been subdivided into two subsections; the typical *Papeda* section and *Papedocitrus*, which have flowers like *Citrus* and leaves like *Papeda*. The vascular anatomy of the flower is simpler than in the subgenus *Citrus*, and is rather similar to the five genera other than *Citrus* included in the true citrus fruits.

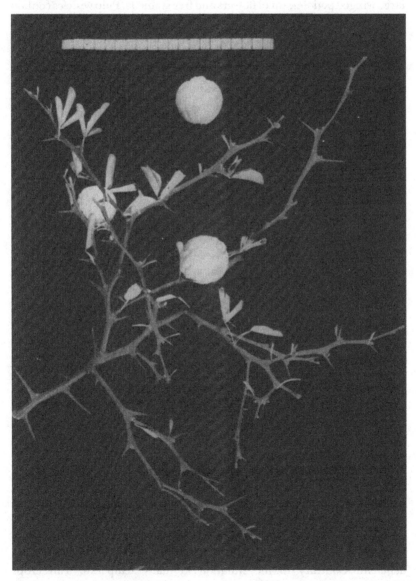

Figure 2.7 *Poncirus trifoliata*

Citrus ichangensis belongs to subsection *Papedocitrus*, and seems to be the most cold resistant of all evergreen species in the orange subfamily. It is characterized by large flowers, connate stamens and extremely large winged petioles. Yuzu (known as *Citrus junos* by Tanaka) is considered to be a hybrid of *C. ichangensis* and a mandarin. It is appreciated in Japan for its acid fruits and as a rootstock for Satsuma. Species of the section *Papeda* have large winged petioles, small flowers and free stamens. Pulp vesicles contain acrid, bitter oil. They are often attached not only to the dorsal walls of the locules, but also to the radial walls for a half to three-quarters of the distance from the dorsal wall to the central core of the fruit, in a manner similar to that found in *Clymenia*. Many species still occur in a truly wild state, in the monsoon region, in contrast to the edible forms of *Citrus*. *Citrus macrophylla*, probably a hybrid of *C. celebica* or some other species of *Papeda*, with possibly *Citrus grandis* (pummelo) as the second parent, is of interest as a rootstock, especially for lemons, conferring early bearing and high tolerance to *Phytophthora* rot. It is sensitive to tristeza virus.

Citrus species

Classification of the subgenus *Citrus* – 16 species according to Swingle (1943) – is still very controversial. It is supposed to include the edible species of *Citrus* (though *C. tachibana* has a bitter juice, and is nearly inedible). Species of the subgenus *Citrus* are characterized by pulp vesicles nearly free from oil droplets, and never containing acrid oil; the petioles have narrow wings or are wingless and if broadly winged, subcordate and less than three quarters of the width of leaf blades. The flowers are large (2.5–4.5 cm diameter) and fragrant, with stamens clustering in bundles. The list of species of *Citrus*, according to Swingle (1943) and Barrett and Rhodes (1976), is given in Table 2.2. Only eight out of the ten species of the subgenus *Citrus* were part of the Barrett and Rhodes study.

The genus *Citrus*, and especially the edible species and cultivars, has undergone a very long period of cultivation. It is very difficult to ascertain the centre of origin of most citrus species as they have been subjected to natural hybridization and probably also cultivation since ancient times. The classification of *Citrus* has proven particularly difficult due to several factors: comparative ease of hybridization, production of adventive embryos (nucellar polyembryony), obliteration of most of the original habitats, the presence of numerous cultivars and hybrids, including spontaneous mutants, and, in some cases, inadequate descriptions and specimens.

Moreover, our knowledge of the wild progenitors of *Citrus* is very deficient. Contrary to the situation in many temperate fruit crops (Rosaceous fruit trees, *Vitis*), where the wild relatives and progenitors

Table 2.2 *List of species of* Citrus[1] *with conventional names, suggested origin[2]
and species concept[3]*

Species	Year named	Conventional name	Assumed native habitat	Species concept
Subgenus *Citrus*				
C. medica L.	1753	Citron	India	True species
C. aurantium L.	1753	Sour orange	China	Hybrid origin (C. reticulata × C. grandis)
C. sinensis Osbeck	1757	Sweet orange	China	Hybrid origin (C. reticulata × C. grandis)
C. grandis Osbeck	1765	Pummelo	China	True species
C. limon (L.) Burm.f.	1766	Lemon	India	Hybrid origin (trihybrid involving C. medica, C. grandis and Microcitrus)
C. reticulata Blanco	1837	Mandarin	China	True species
C. aurantifolia Christm.	1913	Common lime	Malaya	Hybrid origin (trihybrid parentage similar to lemon)
C. paradisi Macf.	1930	Grapefruit		Hybrid origin (C. grandis × C. sinensis)
C. tachibana Tan.	1924	Tachibana	Japan	—[4]
C. indica Tan.	1931	Indian wild orange	India	—[4]
Subgenus Papeda				
C. hystrix D.C.	1813	Mauritius papeda	S.E. Asia	—[4]
C. macroptera Mont.	1860	Melanesian papeda	S.E. Asia	—[4]
C. celebica Koord.	1898	Celebes papeda	Celebes	—[4]
C. ichangensis Swing.	1913	Ichang papeda	China	—[4]
C. micrantha Webster	1915	Papeda	Philippines	—[4]
C. latipes	1928	Khasi papeda	Assam	—[4]

[1] According to Swingle (1943).
[2] Based on Cooper and Chapot (1977).
[3] Based on Barrett and Rhodes (1976).
[4] Not investigated.

seem to have been identified, the taxonomy of the wild members of
subgenus *Citrus* has not yet been satisfactorily worked out. We have only
fragmentary information on wild types in this subgenus. It is extremely
difficult to delimit species and intraspecific taxa in the wild gene pool. We
have also to bear in mind that no barrier of sterility exists within the
subgenus *Citrus*, defined formerly by Swingle (1943) as *Eucitrus*.

Even assuming independent starts from three or more different wild types, the expansion of citrus cultivation has brought these together, resulting in repeated hybridization between types, and the formation of a large complex, obscuring the possibility of clear-cut species delimitation.

In addition, the taxonomic work is complicated by the fact that in *Citrus* sexual variants are represented by one plant, while asexual ones are represented by many identical plants. *Citrus* often shows a very large degree of variation, and abundant natural crossing has given rise to a remarkable degree of heterozygosity, although the free exchange of genes has very often been prevented by the widespread phenomenon of apomixis. The accepted biological concept of species postulates the existence of barriers to the exchange of genes between different species, and formation of species with the aid of such barriers. Such a concept cannot be easily reconciled with the presence of agamic complexes. Stebbins (1950), discussing the species concept in agamic complexes, stated that free exchange of genes between apomicts is prevented by the very nature of their type of reproduction, while the origin of many apomictic clones is from genotypes which have combined the genes of previously isolated sexual species, and which without apomixis would not be able to persist in nature because of sexual sterility. Thus, systematists have not been able to agree on the species boundaries in the apomictic genera. A further phenomenon complicating taxonomic delimitation in *Citrus*, but also of practical significance, is the rejuvenation by neophyosis (Swingle, 1932) of nucellar-bud embryos of more-or-less senescent cultivars that are propagated asexually. Such rejuvenated nucellar 'strains' of old cultivars might be mistaken for taxonomically different cultivars or even subspecies, because of their large leaves and fruit, and highly developed spines. Two widely differing systems of classification, those of Tanaka (1954) and of Swingle (1943), as well as several variants, have been published. Much earlier, Linnaeus (1753) based *Citrus* taxonomy on garden cultivar forms known at his time in Europe: citron, lime, lemon, sweet orange and sour orange. In the revised account of Engler of *Rutaceae* (1931) we find 11 species in the genus *Citrus* and six in the two genera *Poncirus* and *Microcitrus*. Tanaka (1954) considered there to be 145 species, according species rank to numerous cultivars and presumable hybrids, and gave 157 species in a later publication (1961). Until recently, the Swingle system (1943), fully expounded by Swingle and Reece (1967), has been accepted by most authorities as the most valid biological concept of classification. Swingle (1943) first mentioned the possibility of using glycosides as a taxonomic marker, in addition to morphology and other considerations. An expanded list of species, compared with Swingle's classification, was advocated by Hodgson (1961), increasing the number of species from 16 to 31 by including, among others, Rough lemon

(*C. jambhiri*) and sweet lime (*C. limettoides*), and by subdividing *C. reticulata* into groups.

Although the information on wild citrus closely related to the cultivated forms is still very fragmentary, three main groups (species) have been recognized in *Eucitrus*: *Citrus grandis* (*C. maxima*), *Citrus medica* and *Citrus reticulata*. Each might be possibly found to be represented by some wild derivatives, as well as by cultivated varieties.

A new exhaustive study on affinity relationships in cultivated citrus and close relatives by Barrett and Rhodes (1976), using 146 characters of tree, leaf, flower and fruit, points to only three true species in *Eucitrus*: citron (*C. medica*), pummelo (*C. maxima*, formerly *C. grandis*) and *C. reticulata*. Similar conclusions have been communicated by Scora (1975). The first two species are strictly monoembryonic with only sexual offspring. In *C. reticulata* both monoembryony and polyembryony are widespread. Stone *et al.* (1973) report on a new species, *C. halimii*, from Malaya and peninsular Thailand. From China, which is still potentially rich in *Citrus* resources, *C. daoxiamensis* and *C. mangshanensis*, claimed as species, have been recently described (Liu Gengfeng *et al.*, 1990). Barrett and Rhodes (1976) postulate relationships indicating a probable origin of cultivated *Citrus*, and in a few cases, between *Citrus* and related genera. Some of the affinity relationships postulated are shown in Figure 2.8. Citrus biotype interrelationships have been recently reviewed by Scora (1988).

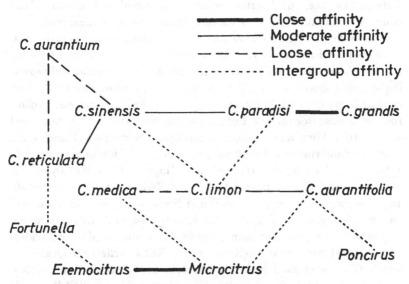

Figure 2.8 Affinity relationships between some *Citrus* species and relatives, according to the Swingle classification (1943) and Barrett and Rhodes (1976)

Both sweet orange and sour orange (as well as the Temple cultivar) are considered by Barrett and Rhodes (1976) to be of *C. reticulata* genotype introgressed with genes from *C. grandis*, now called *C. maxima* (Burm.) Merr. (Scora and Nicolson, 1986). Tatum *et al.* (1974) present chemical evidence that sweet orange (*C. sinensis*) has characteristics of both *C. reticulata* and *C. maxima*. Grapefruit (*C. paradisi*, now also classified as *C. maxima* var. *racemosa*) is presumably a mutant from pummelo or rather a hybrid of pummelo (*C. maxima*). The opinion that grapefruit might have arisen as a hybrid of pummelo and sweet orange has been put forward by Swingle (1943), Robinson (1952), Albach and Redman (1969), Barrett and Rhodes (1976) and Scora *et al.* (1982). Hybrids resembling grapefruit in most aspects have been obtained by Spiegel-Roy and Vardi (unpublished) by crossing pummelo and Temple cv. Grapefruit has never been found native in Asia. A study of the Caribbean 'forbidden fruit', known as 'Shaddette' in the West Indies, has been recently made (Bowman and Gmitter, 1990). While grapefruit may have been described by Hughes in 1750 the name itself was mentioned first at a later date, 1818 (Bowman and Gmitter, 1990).

Species of *Eucitrus* described by Swingle and Reece (1967) include the monoembryonic *C. indica*, which possesses the characters of a wild species prevalent in Assam, though the possibility of it being a hybrid cannot be excluded. *C. tachibana*, found in Southern Japan, Ryukyu Islands and Taiwan (Tanaka, 1922), rather resembles *C. reticulata*. It does not bear edible fruit, but it has been part of some of the oldest Japanese traditions.

Since Swingle's classification, numerous studies have been carried out, employing various techniques in an attempt to help to shed light on species relationships. These include polyphenol oxidase catalyzed browning of young shoots (Esen and Soost, 1978) and seed teguments (Gogorcena and Ortiz, 1988), branching in juice vesicles (Tisserat *et al.*, 1988), root peroxidase isoenzymes (Button *et al.*, 1976), leaf isozymes (Soost and Torres, 1982; Hirai *et al.*, 1986), fraction I protein in leaves (Handa *et al.*, 1986), leaf and rind oils (Malik *et al.*, 1974), essential leaf oils (Scora *et al.*, 1969), long chain hydrocarbon profiles (Nagy and Nordby, 1972), the already mentioned flavonoids (Albach and Redman, 1969; Tatum *et al.*, 1974), limonoids in seeds (Rouseff and Nagy, 1982) and polymethoxyflavones (Mizuno *et al.*, 1991). More recently, use of RFLPs (restriction fragment length polymorphisms) has further contributed to our understanding and interpretation (Roose, 1988). RFLPs reflect the location at which DNA is cleaved by restriction enzymes (Botstein *et al.*, 1980). *C. medica* has been found relatively monomorphic for RFLPs, while *C. reticulata* and *C. maxima* have been shown to be quite polymorphic (Roose, 1988). Also, other taxa have high (>50%) heterozygosity for

RFLPs, supporting the contention that they originated by hybridization. In *C. tachibana* and *C. indica*, unique alleles appear at one or more RFLP loci (Roose, 1988).

RFLP and RAPD (random amplified polymorphic DNA) can detect differences at the DNA level. RAPD is a DNA polymorphism assay based on the amplification of random DNA segments with single primers of arbitrary nucleotide sequence (Williams *et al.*, 1990). Polygenic relationships in citrus as revealed by diversity of cytoplasmic genomes have been also studied by RFLPs, using nine mitochondrial and one chloroplast DNA probe/restriction enzyme combinations (Yamamoto *et al.*, 1993). DNA polymorphisms among mandarins were investigated by Omura *et al.* (1993), using the polymerase chain reaction (PCR) developed by Saiki *et al.* (1988). DNA fingerprinting has been used to distinguish between citrus cultivars (Matsuyama *et al.*, 1993) and within *Poncirus* strains (Komatsu *et al.*, 1993).

Studies on the plastome of *Citrus* (Green *et al.*, 1986) indicate no differences between plastomes of *C. aurantium*, *C. grandis* (*C. maxima*), *C. paradisi* and *C. limon*. This would point to the monoembryonic, multi-seeded *C. maxima* as the female parent of sour orange, grapefruit and lemon. *C. reticulata* and *C. medica*, recognized as true species by Barrett and Rhodes in addition to *C. maxima*, gave distinct plastome restriction patterns. *C. limon* has been considered (Barrett and Rhodes, 1976) (Figure 2.8) to be a trihybrid involving *C. maxima*, *C. medica* and possibly *Microcitrus*. However, as all three postulated parents are monoembryonic, origin of the polyembryonic lemon may be still unsolved. Citron and pummelo have been found to contribute to its origin (Malik *et al.*, 1974; Green *et al.*, 1986).

The chapter on *Citrus* and its relatives will be incomplete without considering an additional subtribe, seemingly not closely related to the other subtribes of Aurantioideae. The Balsamocitrinae, the hard-shelled citroid fruit trees, have no pulp vesicles, but a hard, woody exocarp. They have been considered a side branch paralleling Citrinae. Of the seven genera belonging to the subtribe, four are found in south-eastern Asia from India to Burma and Indochina, and three in tropical Africa. Their ovaries have 6–20 locules with 6–16 ovules in each locule. The locules are filled with a resinous gum. The seeds are wooly, covered with hair. The subtribe has been divided into three groups with the following genera, A: *Swinglea*, B: *Aegle* (Figure 2.9), *Balsamocitrus*, *Afraegle*, *Aeglopsis*, C: *Feronia*, *Feroniella*. It has been speculated that *Balsamocitrinae* and *Citrinae* may have evolved from a common ancestral form, but this has not been substantiated. *Feronia limonia*, which is deciduous, is also grown for its fruit, for jelly and chutney. *Citrus* has been grafted onto *Swinglea*, reaching a good size in warm soil. *Aegle marmelos* (Figure 2.9) yields the 'bael fruit' esteemed in India. The twigs are dimorphic. The tree, which has deciduous leaves, has some resistance to cold. *Citrus* has been grafted onto *Feronia* and onto *Feroniella*.

Importance of citrus relatives

Few cultivated crop plants possess such a significant and varied group of wild relatives as *Citrus*: in the genus *Citrus* itself, in closed related genera and in closely related tribes. *Citrus* relatives exhibit a wide range of adaptability to climatic and soil conditions, exemplified by the tolerance to cold of *Poncirus*, to boron by *Severinia*, to *Phytophthora* by *Citropsis*, and to

Figure 2.9 *Aegle marmelos* (courtesy Dr Roose)

the burrowing nematode by *Microcitrus*, and by the adaptation to salt by *Eremocitrus*. While it seems easier to exploit similar traits in the search for further rootstocks than in that for new citrus cultivars, progress in somatic hybridization and the advent of genetic engineering will facilitate the future use of the *Citrus* relatives more than by the use of breeding efforts by hybridization only.

Large areas of the Monsoon region and certain areas of tropical Africa have been inadequately explored for *Citrus* relatives. As plants become adapted to special environmental conditions over a very long period, it is reasonable to expect that certain wild relatives of *Citrus* will possess valuable physiological peculiarities. If successfully conserved, wild relatives will not only be used in future breeding efforts aimed at solving problems as yet unrecognized, but will also be used as study and research material for elucidating some of the difficulties encountered in the phylogeny and taxonomy of *Citrus*.

Horticultural classification of cultivated citrus

All commercially used scion and rootstock cultivars belong to the genus *Citrus*, except kumquats, *Fortunella* spp., and trifoliate orange, *Poncirus trifoliata*, which is used as a rootstock only.

Originally, citrus trees were grown as seedlings, and in some cases from air layers, but today most cultivars are budded onto rootstocks (see Chapters 5 and 6). Rootstocks affect yield and fruit quality as well as tolerance to biotic (viruses, pests and diseases) and abiotic (cold, drought, salt) stress. Choice of rootstock will also depend on compatibility with particular scion cultivars.

Sweet oranges (*Citrus sinensis*) can be placed into four fairly distinctive groups.

1   Navel oranges. This is the most important group for fresh fruit. Navel oranges have the prominent distinctive feature of a small, 'secondary' fruit embedded in the apex of the main fruit. Worldwide expansion of this type started after budwood was sent from Bahia, Brazil to the US Department of Agriculture in 1870. Since then numerous early- and late-ripening mutants of Navel have been discovered and propagated.

2   Common oranges (also known as 'blond' oranges). These include Valencia, used for fresh fruit and processing, Shamouti, with its typical form and flavour, and Pera, Hamlin and Pineapple, grown mainly for processing.

3 Pigmented oranges, with anthocyanin in the rind and juice, known
   as 'blood oranges'. The best-known are Moro, Tarocco, and
   Sanguinelli.
4 Acidless or sugar oranges, with very low acidity in the fruit (about
   0.2%), called Sukkari in Egypt, de Nice in France.

There are several putative natural hybrids between orange and mandarin (*C. reticulata*). Their origin has not been ascertained. Some of the
best known are Temple orange, Iyokan, Ortanique. Similar types have
been produced for processing by hybridization, with an emphasis on
orange characteristics.

Sour orange (*C. aurantium*) is still widely used as a rootstock, in spite of the
high sensitivity of trees grafted on it to tristeza virus. Sour oranges are
highly bitter, and contain neohesperidin. Sevillano sour orange fruit is
mainly used for processing into marmalade. Flowers of sour orange and of
special selections are used for the production of oil of neroli (for perfumes).
A hybrid of sour orange, bergamot, is grown for its distinctively perfumed
oil and for scenting tea. Special sour orange types, such as the Chinotto,
are grown as ornamentals.

Mandarins (mainly *C. reticulata*) probably originated in south-west
China, and they may have been cultivated in China for several thousand
years. From the tenth century AD they were widely cultivated in Japan.
Cultivation on a world scale started only in the nineteenth and twentieth
century, achieving great prominence in Spain. Satsuma, from Japan, and
Ponkan, from several places in Asia, were widely adopted considerably
earlier. Mandarins, in Spain, have attained prominence since the 1950s.
The name tangerine has often been used synonymously with mandarin,
especially in the USA. Mandarins are a most varied group and certain
authors have assigned species names to the different groups. They have
been classified by Hodgson into five groups:

1 Satsuma (unshu mikan), also classified as *Citrus unshiu*.
2 Mediterranean mandarin, also classified as *C. deliciosa*.
3 King mandarin (*C. nobilis* Loureiro).
4 Common mandarins (*C. reticulata*).
5 Small-fruited mandarins.

Numerous mutants of Satsuma, some of them of nucellar origin, have
been discovered in Japan and Spain. Numerous mutants of Clementine,
all of them bud mutants, have also been discovered and propagated.

Some explanations have been offered on the origin of the Clementine, a
most widely grown seedless mandarin. According to one, it resulted from
a cross between the Mediterranean mandarin and pollen from an ornamental sour orange (Granito), discovered in Algeria around 1890. Some

authorities believe it to be identical to a variety known as the Canton mandarin in China. Samaan (1982) points to Baladi mandarin or possibly Baladi Blood orange as its seed parent.

Clementine is self-incompatible, and in the presence of pollen donors from other clones the fruit becomes very seedy. It has, however, considerable parthenocarpic capacity.

A widely grown mandarin is Ponkan, of very large significance in India, China, the Philippines and Brazil.

A fairly large group of mandarin hybrids has been found and propagated, and also produced by breeding. These are treated in commerce in a manner similar to mandarin or 'easy peeling fruit'. Those originating from a cross between grapefruit and mandarin are called 'tangelos' (Minneola, Orlando). Some are presumed hybrids between mandarin and sweet orange (tangors) and some cultivars widely grown in Japan may have originated as hybrids between pummelo (*C. maxima*) and *C. reticulata*.

Several mandarin type cultivars, from breeding, are backcrosses of tangelos (grapefruit × mandarin hybrids) to mandarins, mostly Clementine.

Grapefruit probably originated from a natural cross between pummelo as a seed parent and sweet orange or some other similar parent as pollen donor. It has achieved prominence in the twentieth century as fresh fruit and for processing. It is not clear whether the name was given because the flavour resembles the grape or, more probably, because the fruits are borne in clusters, contrasting with the single-borne fruit of the pummelo. Pigmented grapefruits deriving their colour from lycopene have become prominent and are now often preferred to white skinned and fleshed grapefruit for fresh fruit. Pigmented varieties originated in Texas and Florida. However, Marsh Seedless (white) is still the most important grapefruit universally. Pigmented cultivars such as Ray Ruby, Star Ruby and, recently, Rio Red increase in importance (Saunt, 1990).

The *pummelo*, sometimes referred to as 'shaddock', is a typically large sized tropical citrus fruit. It is either grouped into two classes, white and pigmented, or according to the country where the cultivars were developed – Thailand, China and Indonesia. Pummeloes show an enormous variation in fruit size. Low-acid pummeloes, similar to low-acid oranges, also exist and have proven to be most valuable germ plasm. Pummelo is self incompatible and hybridizes readily, and many different varieties have arisen.

A new type of fruit has resulted from a cross between acidless pummelo and grapefruit (Soost and Cameron, 1980, 1985). The varieties are called 'Oroblanco' and 'Melogold' and their fruit has low acidity.

Table 2.3 *Some well known cultivars of Citrus*

Orange	Mandarin	Mandarin type fruit, tangor	Tangelos (grapefruit × mandarin)	Grapefruit	Pummelo	Lemon	Kumquat
Bahianinha	Avana	Ellendale	Orlando	Duncan	Banpeiyu	Eureka	Marumi
Hamlin	Clementine	Iyokan	Minneola	Marsh seedless	Chandler	Femminello	Meiwa
Maltaise demi-sanguine	Cravo	King	Seminole	Star Ruby	Goliath	Fino	Nagami
Moro	Dancy	Kiyomi	Natsudaidai (pummelo × mandarin)		Hirado	Genoa	
Navel	Emperor	Murcott	Hassaku (pummelo × tangelo)		Kae Panne	Interdonato	
Pera	Encore	Ortanique			Kao Phuang	Lamas	
Pineapple	Fortune	Temple	Oroblanco (pummelo × grapefruit)		Mato	Lapithkiotiki	
Sanguinelli	Fairchild		Melogold		Pandan Bangi	Lisbon	
Shamouti	Fremont				Shatinyu	Monachello	
Succari	Imperial				Thong Dee	Villafranca	
Tarocco	Kinnow						
Tomango	Nova						
Valencia	Ponkan						
	Satsuma						
	Tankan						

Numerous mutants and nucellar selections have assumed great significance and expansion; best known are Atwood, Fisher, Navelina, Newhall, Navelate, Palmer, Summerfield, Lane's Navel, Leng (from Navel); Marisol, Arrufatina, Oroval, Nules, Hernandina, Nour (from Clementine); Miyamoto Wase, Ueno Wase, Yamakana Wase, Clausellina, Miyagawa Wase, Okitsu Wase, Hayashi Unshu, Owari Unshu, Aoshima Unshu (from Satsuma); Ruby Red (from pink Marsh grapefruit); Henderson, Ray Ruby, Rio Red (from Ruby Red), Flame (from Henderson).

Lemon has been considered a trihybrid (Barrett and Rhodes, 1976). It has not been found growing wild, but similar natural hybrids are indigenous to Punjab. It is rather difficult to identify fruit of a particular cultivar with certainty because the range of fruit shape from different 'flushes' of growth of the same cultivar varies often as much as between cultivars.

A distinction is made between lemon cultivars such as Eureka and Lisbon, and the Italian lemon cultivars (mainly Femminello). The main Spanish lemons are Fino and Verna. Certain lemons may be hybrids between lemon and citron. Another lemon (Meyer) is low acid and rather resembles an orange in shape.

Limes include both acid and sweet ('acidless') cultivars. Sour limes consist of small-fruited Indian, West Indian or Mexican lime (*C. aurantifolia*) and large-fruited Tahiti or Persian lime (*C. latifolia*). Bigeneric cultivars (lime × kumquat) are also known.

The *citron* (*Citrus medica*) was considered to be the only citrus fruit known to the Greeks, Hebrews and Romans. Citron fruits are highly variable in size, and the fruits often have a persistent style. They are used in rituals and in the production of candied peel. For Jewish religious rituals the fruit has to be unblemished, with a persistent style and borne on a tree raised from seed or a cutting.

A few cultivars of the cold-hardy *Fortunella* (kumquat) are known, while one cultivar is grown as an ornamental. Limequats resulted from crosses between the sour Mexican lime and kumquat. Orangequats and trigeneric citrangequats are also known. A list of some well-known cultivars of *Citrus* is given in Table 2.3.

The trifoliate orange (*Poncirus*) is important as a rootstock. Some of its bigeneric hybrids, citranges (*Poncirus* × orange), and citrumelo (*Poncirus* × grapefruit), are valuable rootstocks. Combinations approaching edibility have been evolved from crosses of *Poncirus* × grapefruit F_1 with orange (Barrett, 1990).

Recommended reading

Hodgson, R. W. (1967). Horticultural varieties of citrus. In *The Citrus Industry*, Vol. 1, ed. W. Reuther, H. J. Webber and L. D. Batchelor, pp. 431–591. Berkeley: Division of Agricultural Sciences, University of California.

Saunt, J. (1990). *Citrus Varieties of the World*. Norwich, England: Sinclair International. 126 pp.

Swingle, W. T. and Reece, P. C. (1967). The botany of *Citrus* and its wild relatives. In *The Citrus Industry*, Vol. 1, ed. W. Reuther, H. J. Webber and L. D. Batchelor, pp. 190–430. Berkeley: Division of Agricultural Sciences, University of California.

Sykes, S. R. (1989). Overview of the family *Rutaceae*. In *Citrus Breeding Workshop*, ed. R. R. Walker, pp. 93–100. Adelaide, Australia: CSIRO.

Tanaka, T. (1954). *Species Problem in Citrus* (Revisio aurantiacearum IX) *Jap. Soc. Prom. Sci.* Tokyo: Ueno. 152 pp.

Literature cited

Albach, R. F. and Redman, G. H. (1969). Composition and inheritance of flavanones in citrus fruit. *Phytochemistry*, **8**: 127–43.

Barrett, H. C. (1978). Intergeneric hybridization of *Citrus* and other genera in citrus variety improvement. *Proc. Int. Soc. Citriculture* 1977, ed. W. Grierson, pp. 586–9. Lake Alfred, FL: ISC.

Barrett, H. C. (1990). US 119 an intergeneric hybrid citrus scion breeding line. *HortScience*, **25**: 1670–1.

Barrett, H. C. and Rhodes, A. M. (1976). A numerical taxonomic study of affinity relationships in cultivated *Citrus* and its close relations. *Syst. Bot.*, **1**: 105–36.

Bitters, W. P., Cole, D. A. and Bruscan, A. (1969). The citrus relatives as citrus rootstocks. *Proc. First Int. Citrus Symp.*, ed. H. D. Chapman, Vol. 1, pp. 411–15. Riverside: University of California.

Botstein, D., White, R. L., Skolnick, M. H. and Davis, R. W. (1980). Construction of a genetic map in man using restriction fragment length polymorphism. *Am. J. Hum. Genet.*, **32**: 314–31.

Bowman, K. D. and Gmitter, F. G. Jr (1990). Caribbean forbidden fruit: Grapefruit's missing link with the past and bridge to the future? *Fruit Var. J.*, **44**: 41–4.

Button, J., Vardi, A. and Spiegel-Roy, P. (1976). Root peroxidase isoenzymes as an aid in *Citrus* breeding and taxonomy. *Theor. Appl. Genet.*, **47**: 119–23.

Cooper, W. C. and Chapot, H. (1977). Fruit production – with special emphasis on fruit for processing. In *Citrus Science and Technology*, Vol. 2, ed. A. Nagy, P. E. Shaw and M. K. Veldhuis, pp. 1–24. Westport, CT: Tel Avi Publ. Co.

Engler, A. (1896). Rutaceae. In *Die naturlichen Pflanzenfamilien*, ed. A. Engler and K. Prantl, Vol. 3, No. 4, pp. 95–201. Leipzig: Engelmann.

Engler, A. (1931). Rutaceae. In A. Engler and K. Prantl. *Die naturlichen Pflanzenfamilien*, 2nd edn., pp. 187–359. Leipzig: Engelmann.

Esen, A. and Soost, R. K. (1978). Separation of nucellar and zygotic citrus seedlings by use of polyphenol oxidase-catalyzed browning. *Proc. Int. Soc. Citriculture* 1977, ed. W. Grierson, Vol. 2, pp. 616–18. Orlando, FL: ISC.

Gogorcena, Y. and Ortiz, J. M. (1988). Seed teguments as an aid in citrus chemotaxonomy. *J. Hort. Sci.*, **63**: 687–94.

Green, R. M., Vardi, A. and Galun, E. (1986). The plastome of *Citrus*. Physical map variation among *Citrus* cultivars and species, and comparison with related genera. *Theor. Appl. Genet.*, **72**: 170–7.

Handa, T., Ishizawa, Y. and Oogaki, C. (1986). Phylogenetic study of fraction I protein in the genus *Citrus* and its close related genera. *Japan J. Genet.*, **61**; 15–24.

Hirai, M., Kozaki, I. and Kajiura, I. (1986). Isozyme analysis and phylogenetic relationship of citrus. *Japan J. Breed.*, **36**: 377–89.

Hodgson, R. W. (1961). Taxonomy and nomenclature in Citrus. *Int. Org. Citrus Virol. Proc.*, **2**: 1–7.

Jones, D. T. (1989). Progress in taxonomic research on the Aurantioideae of Southeast Asia. Paper presented at the *FAO-UNDP Asian Citrus Rehabilitation Conference*, July 1989, Malang, Indonesia.

Komatsu, A., Akihama, T., Hidaka, T. and Omura, M. (1993). Identification of Poncirus strains by DNA fingerprinting. In *Techniques on Gene Diagnosis and Breeding by Fruit Trees*, ed. T. Hayashi, M. Omura and N. S. Scott, pp. 88–95. Japan: FTRS.

Linnaeus, C. (1753). *Species Plantarum*. 2 vols. Stockholm.

Liu Gengfeng, He Shanwen and Li Wenbin (1990). Two new species of Citrus in China. *Acta Bot. Yunnanica*, **12**: 287–9.

Malik, M. N., Scora, R. W. and Soost, R. K. (1974). Studies on the origin of the lemon. *Hilgardia*, **42**: 361–82.

Matsuyama, T., Omura, M. and Akihama, T. (1993). DNA fingerprinting in *Citrus* cultivars. In *Techniques on Gene Diagnosis of Breeding Fruit Trees*, ed. T. Hayashi, M. Omura and N. S. Scott, pp. 26–30. Japan: FTRS.

Mizuno, M., Linuma M., Ohara, M., Tanaka, T. and Iwamasa, M. (1991). Chemotaxonomy of the genes *Citrus* based on polymethoxyflavones. *Chem. Pharm. Bull.*, **39**: 945–9.

Nagy, S. and Nordby, H. E. (1972). Saturated and mono-unsaturated long-chain hydrocarbon profiles of lipids from orange, grapefruit, mandarin and lemon juice sacs. *Lipids*, **7**: 666–70.

Omura, M., Hidaka, T., Nesumi, H., Yoshida T. and Nakamura, I. (1993). PCR markers for *Citrus* identification and mapping. In *Techniques on Gene Diagnosis and Breeding in Fruit Trees*, ed. T. Hayashi, M. Omura and N. S. Scott, pp. 66–73. Japan: FTRS.

Robinson, T. R. (1952). Grapefruit and pummelo. *Econ. Bot.*, **6**., 228–45.

Roose, M. L. (1988). Isozymes and DNA restriction fragment length polymorphisms in a citrus breeding and systematics. In *Proc. Sixth Int. Citrus Congr.*, ed. R. Goren and K. Mendel, Vol. I, pp. 155–65. Philadelphia/Rehovot: Balaban Publishers; Weikersheim, Germany: Margraf Scientific Books.

Rouseff, R. L. and Nagy, S. (1982). Distribution of limonoids in *Citrus* seeds. *Phytochemistry*, **21**: 85–90.

Saiki, R. K., Gelfand, D. H., Staffel, S., Scharf, S. J., Higuchi, R., Horn, G. T., Mullis, K. B. and Erlich, H. A. (1988). Primer directed amplification of DNA with a thermostable DNA polymerase. *Science*, **239**: 487–91.

Samaan, L. G. (1982). Studies on the origin of Clementine tangerine (*Citrus reticulata Blanco*). *Euphytica*, **31**: 167–73.

Saunt, J. (1990). *Citrus Varieties of the World*. Norwich, England: Sinclair International. 126 pp.

Scora, R. W. (1975). On the history and origin of citrus. *Bull. Torrey Bot. Club*, **102**: 369–75.

Scora, R. W. (1988). Biochemistry, taxonomy and evolution of modern cultivated citrus. In *Proc. Sixth Int. Citrus Congr.*, ed. R. Goren and K.

Mendel, Vol. I, pp. 277–89. Philadelphia/Rehovot: Balaban Publishers; Weikersheim, Germany: Margraf Scientific Books.

Scora, R. W. and Nicolson, D. H. (1986). The correct name for the Shaddock, *Citrus maxima*, not *C. grandis* (*Rutaceae*). *Taxon*, **35**: 592–5.

Scora, R. W., Duesch, G. and England, A. B. (1969). Essential leaf oils in representatives of the *Aurantioideae* (*Rutaceae*). *Am. J. Bot.*, **56**: 1094–102.

Scora, R. W., Kumamto, J., Soost, R. K. and Nauer, M. (1982). Contribution to the origin of the grapefruit, *Citrus paradisi* (Rutaceae). *Syst. Bot.*, **7**: 170–7.

Soost, R. K. and Cameron, J. W. (1980). Oroblanco a triploid pummelo – grapefruit hybrid. *HortScience*, **15**: 667–9.

Soost, R. K. and Cameron, J. W. (1985). Melogold a triploid pummelo–grapefruit hybrid. *HortScience*, **20**: 1134–5.

Soost, R. K. and Torres, A. M. (1982). Leaf isozymes as genetic markers in *Citrus*. In *Proc. Int. Soc. Citriculture 1981*, ed. K. Matsumoto, pp. 7–10. Okitsu, Shizuoka, Japan: Okitsu Fruit Tree Research Station.

Stebbins, G. K. Jr (1950). *Variation and Evolution in Plants*. New York: Columbia University Press. 643 pp.

Stone, B. C., Lowry, J. B., Scora, R. W. and Kwiton, J. (1973). *Citrus halimii*: a new species from Malaya and Peninsular Thailand. *Biotropica*, **5**: 102–10.

Swingle, W. T. (1932). Neophyosis or rejuvenescence of nucellar bud seedlings in Citrus. *Am. J. Bot.*, **19**: 839. (Abstract.)

Swingle, W. T. (1943). The botany of *Citrus* and its wild relatives of orange subfamily (family Rutaceae, subfamily Aurantioideae). In *The Citrus Industry*, Vol. I, ed. H. J. Webber and L. D. Batchelor, pp. 128–474. Berkeley: Division of Agricultural Sciences, University of California.

Swingle, W. T. and Reece, P. C. (1967). The botany of *Citrus* and its wild relatives. In *The Citrus Industry I*, ed. W. Reuther, H. J. Webber and L. D. Batchelor, pp. 190–430. Berkeley: Division of Agricultural Sciences, University of California.

Sykes, S. R. (1993). Evaluation of germplasm for *Citrus* breeding in Australia. In *Techniques on Gene Diagnosis and Breeding in Fruit Trees*, ed. T. Hayashi, M. Omura and N. S. Scott, pp. 219–30. Japan: FTRS.

Tanaka, T. (1922). Citrus fruits of Japan; with notes on their history and the origin of varieties through bud variation. *J. Hered.*, **13**: 243–53.

Tanaka, T. (1954). *Species Problem in Citrus* (Revisio aurantiacearum IX) *Japan Soc. Prom. Sci.* Tokyo: Ueno. 152 pp.

Tanaka, T. (1961). *Citrologia Semi Centennial Commemoration Papers on Citrus Studies*. Osaka: Citrologia Supporting Foundation. 114 pp.

Tatum, J. H., Berry, R. E. and Hearn, C. J. (1974). Characterization of *Citrus* cultivars and separation of nucellar and zygotic seedlings by thin-layer chromatography. *Proc. Fla. State Hort. Soc.*, **87**: 75–81.

Tisserat, B., Jones, D. and Galletta, P. D. (1988). Natural branching in *Citrus* juice vesicles. *J. Am. Soc. Hort. Sci.*, **113**: 957–60.

Williams, J. G. K., Kubelik, A. R., Livak, K. J., Rafalski, J. A. and Tingey, S. V. (1990). DNA polymorphisms amplified by arbitrary primers are useful as genetic markers. *Nucleic Acid. Res.*, **18**: 6531–5.

Yamamoto, M., Kobayashi, S., Nakamura, Y. and Yamada, Y. (1993). Phylogenic relationships of *Citrus* revealed by diversity of cytoplasmic genomes. In *Techniques on Gene Diagnosis and Breeding in Fruit Trees*, ed. T. Hayashi, M. Omura and N. S. Scott, pp. 39–46. Japan: FTRS.

3

The vegetative *Citrus* tree:
development and function

Introduction

FOR A COMPREHENSIVE description of the vegetative *Citrus* tree it is not enough to provide a structural description of the basic organ units. Attention must also be paid to physiological activity, at the organ level as well as at the whole-plant level. Environmental, eco-physiological aspects of tree activity must also be considered if a broader understanding of citrus tree behavior is to be achieved.

Citrus trees belong to the 'evergreens', which do not shed their leaves during the fall. The evergreen habit has important consequences for leaf longevity and physiological activity, which must be reflected in leaf design and structure. In the absence of fall abscission the longevity of leaves may extend to a whole year and beyond. The year-round presence of leaves enables uninterrupted, day-by-day photosynthetic activity, and thus a continuous supply of photosynthates. However, in subtropical climate zones winter temperatures might be quite low so that photosynthetic gains during winter are lowered. The evergreen habit may have significant implications with regard to the role of nutrient reserves, particularly during springtime. Whereas deciduous trees are totally dependent for early spring growth upon their carbohydrate reserves, evergreens like citrus may at least partly rely on the supply of photosynthate from the previous season's foliage. The evergreen character also has far-reaching consequences for the annual cycle of flowering and fruiting, as will be discussed in Chapter 4. While keeping these traits in mind, we may now proceed with a discussion of the structure and function of citrus tree organs.

Seed and seedling

The mature citrus seed consists of one or more embryos, enveloped by two seed coats. The outer seed coat (testa) is tough and woody, grayish-white to cream. In some cultivars it extends beyond the embryo to form a beak-like structure (Figure 3.1). The inner seed coat (tegmen) is a thin membrane, formed essentially from the inner integument of the ovule, and also containing remnants of the nucellus and the endosperm (Schneider, 1968). The tegmen has a characteristic light-brown to reddish-purple color. The chalazal spot has a distinct, slightly different color (Figure 3.1), which may be used as a taxonomic character. Most of the volume of the mature embryo is taken up by the cotyledons, which are creamy-white in most cultivars, and greenish in *C. reticulata* and most of its hybrids (Frost and Soost, 1968). In polyembryonic seeds the cotyledons often vary in size and some of the embryos may be small, with poorly developed cotyledons. The rudimentary plumule and radicle lie between the bases of the cotyledons, close to the micropylar end of the seed through which the radicle will emerge.

Germination of citrus is hypogeous. The radicle produces a strong, fleshy taproot. Secondary roots appear after the taproot has reached 8–10 cm and the first pair of true leaves had developed (Figure 3.2). In cases of more than one embryo present the course of germination may be somewhat obstructed (Figure 3.2).

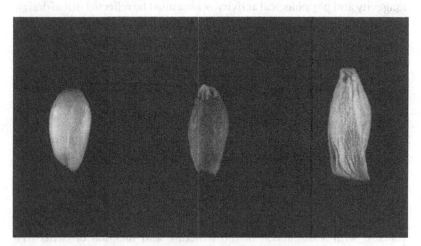

Figure 3.1   Seed of sour orange (*Citrus aurantium*); the chalazal end up, the micropylar down (×2). Right, with both seed coats; note the 'beak' shaped micropylar end of the testa. Middle, without testa; note the color of the tegmen and the distinct chalazal spot. Left, without seed coats, the cotyledons exposed

Figure 3.2 A monoembryonic seedling (right) and three poly-embryonic seedlings (left) of sour orange (*C. aurantium*), two to three weeks after emergence. Note the abnormalities in root development in polyembryonic seedlings

Citrus seeds have a high lipid content and are, therefore, sensitive to excessive drying. About 20 days may elapse between sowing and emergence at an optimum temperature of 30 °C, while up to 80 days may be required at 15 °C (Davies and Albrigo, 1994). Removal of the testa enhances germination considerably.

Sowing is practiced mainly with rootstock cultivars which are grown in nurseries and prepared for grafting. The rate of seedling development varies considerably among cultivars and is greatly dependent upon temperature, soil type, irrigation and, particularly, nitrogen fertilization.

Citrus seedlings are highly juvenile, much more so than rooted cuttings or other vegetatively propagated plants. Juvenility is generally associated with inability to flower, but the juvenile growth habit is revealed also in upright, unbranched growth, abundance of thorns, and in certain cultivars (e.g. Shamouti orange) by very large leaves. It is not uncommon to find in young orchard trees parts of the canopy which are still juvenile while others have already acquired a mature appearance.

Shoot development

Shoot growth occurs in most types of citrus in several well-defined waves (flushes). Under cool climatic conditions only two flushes appear annually, while three to five flushes occur in warmer, subtropical regions. Under wet, tropical conditions shoot growth occurs uninterruptedly, throughout the year, without distinct flushes. Lemons, citrons and acid limes retain their tropical nature even in cooler climates and new shoots emerge year-round. The spring flush is the most important one, containing both vegetative and reproductive shoots (Figure 3.3). The midsummer and subsequent flushes are generally vegetative, with fewer but longer, vigorously growing shoots and larger leaves. As trees get older, the spring flush comprises mainly short, reproductive shoots (leafy and leafless inflorescences, Figure 3.3, see also Chapter 4). For its vegetative growth the tree is dependent upon the summer flushes.

The elongation of vegetative shoots comes to an end through shoot-tip abscission, a phenomenon also known in other tree species (Kozlowski, 1964). Shoot tips about to abscise may be recognized by their light, yellow appearance (Figure 3.4).

Figure 3.3 Drawing of spring flush of Shamouti orange (*C. sinensis*), with vegetative shoot (A), leafy inflorescences (B) and leafless inflorescences (C)

Figure 3.4 Summer flush of Shamouti orange (*C. sinensis*). Note the appearance of the shoot tip, which is on the verge of abscission. Also note the 'ridges' formed along the stem, in a continuation of the leaf petioles

The growth of a new flush originates from an axillary bud close to the top and is, therefore, at a slight angle to the previous one (Figure 3.5). The ensuing sympodial zig-zag pattern is eventually obscured as the stem increases in diameter. Shoot-tip abscission does not normally occur in

Figure 3.5 Emergence of the summer flush in Marsh seedless grapefruit (*C. paradisi*). Note that the new shoot develops at a slight angle to the older shoot

lemon; very long, upright vegetative shoots are not uncommon, therefore, in lemon trees.

The newly forming stem is green and tender, with a prominent ridge extending below the base of each petiole (Figure 3.4). The ridges cause the young stems to be triangular in cross-section (Figure 3.6), but stems eventually become round when secondary thickening takes over. Development of the vascular systems in citrus stems has been described in detail by Schneider (1968). The wood of citrus belongs to the diffuse porous type and is rather dense and hard. Detailed mathematical analysis of the growth of the trunk and branches of grapefruit trees was conducted by Turrell *et al.* (1980).

Leaves are arranged spirally around the stem, and the phyllotaxy of most *Citrus* species, as well as that of *Poncirus* and *Fortunella*, is 3/8. The phyllotaxy of pummelo and grapefruit has been found to be 2/5 (Schneider, 1968). The direction of spirality is reversed with each growth flush (Schroeder, 1953).

An axillary bud occurs in the axil of each citrus leaf (Figure 3.4). The axillary bud consists of an apical meristem, covered by several prophylls (bud scales). Accessory buds develop in the axils of the prophylls; thus, multiple buds are present in the axils of leaves (Figure 3.7). Axillary thorns may subtend the buds, occurring opposite the first prophyll (Figure 3.8). Thorns are particularly prominent in juvenile, vigorously growing shoots.

The apparently simple leaf of citrus is in reality a compound unifoliate leaf. It represents the terminal leaflet of a compound leaf like that of some distant relatives of *Citrus* (e.g. *Citropsis*, Figure 2.2). The leaf of *Poncirus trifoliata* (Figure 2.7) may represent an intermediate phylogenetic stage between a pinnately compound leaf and a unifoliate leaf like that of *Citrus*, where the only trace of the ancestral form is the presence of a joint between the petiole and the blade. In citron (*C. medica*) the petioles are not clearly articulated with the leaf blade.

In most *Citrus* species the petioles are winged (Figure 3.8). Grapefruit and pummelo petiole wings are large, those of sweet orange are smaller and petioles of lemon leaves are without wings. Within each species, larger petiole wings are associated with juvenile growth.

Citrus leaves start their expansion before the termination of stem elongation. The leaf, which is initially light green, reaches 80% of its full size within 1–2 months and then turns dark green and leathery. The dry matter content, which is about 29% in young, fully expanded leaves, increases up to 45% in hardened, 1-year-old leaves. This includes large amounts of cell wall materials as well as starch and other reserve carbohydrates. Calcium oxalate and flavanoids are also regular constituents of

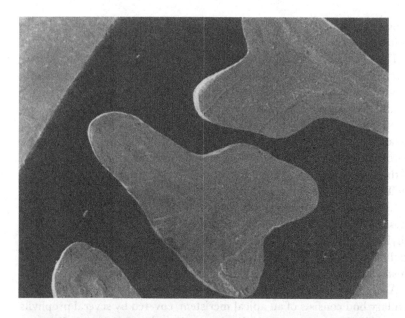

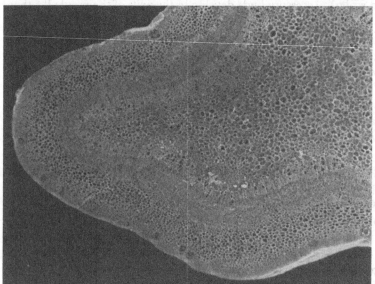

Figure 3.6 Upper, cross section of young, triangular shoot from the summer flush of Shamouti orange (*C. sinensis*) as viewed by SEM. Lower, detail, showing the young cambium tissue between the phloem and the xylem. Note the presence of oil glands along the epidermis

citrus leaves. Stomata are present mainly at the abaxial, lower surface of leaves (400 to 700 per mm^2). Leaves acquire photosynthetic competence rather slowly and start exporting only after all leaves of the shoot are fully expanded. Maximum photosynthetic rates are attained at the age of 3–4 months and some decline in activity has been observed from 6 months on (Fishler, 1985). Leaves remain active, however, until they drop and die, which may take more than two years. Sun and shade leaves differ in

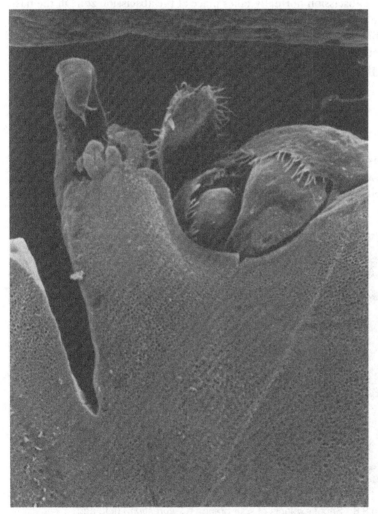

Figure 3.7 A longitudinal section through an axillary bud of Shamouti orange (*C. sinensis*) as viewed by SEM. The major axillary bud is to the left. Accessory buds covered with prophylls are to the right

structure, composition and physiological activity (Monselise, 1951). Syvertsen (1984) has shown, however, that upon transfer to a different light environment, even mature, fully expanded leaves undergo structural and physiological light acclimation changes.

Orange leaf abscission takes place throughout the year. Most intensive drop occurs during the spring blossoming period, including old as well as very young current-season leaves. A second, prolonged period of leaf abscission occurs during the fall (Erickson and Brannaman, 1960).

Leaf abscission normally takes place at the abscission zone in the base of the petiole but under certain types of stress, separation occurs at the junction between the petiole and leaf blade. Hormonal and physiological aspects of citrus leaf abscission have been discussed recently by Goren (1993) (see also fruit abscission in Chapter 4).

Leaf activity

Citrus belong to C_3 plants, with photosynthetic rates lower than those of C_4 plants. Even among the C_3 group citrus are in the low activity range (together with other tropical and subtropical trees), being considerably

Figure 3.8 Shoot explants of sour orange (*C. aurantium*) (left) and Palestine sweet lime (*C. limettoides*) (right) (×0.5). Note the developed petiole wings in sour orange and the absence of wings in sweet lime. Axillary buds and thorns are shown

lower than annual crop plants and lower still compared with deciduous fruit plants such as apple and grape (Kriedemann, 1971).

Assimilation rates of 4 to 8 µmol CO_2 m^{-2} s^{-1} seem realistic under optimal field conditions; higher rates are frequently obtained in greenhouse experiments. Low assimilation rates of citrus are accompanied by low rates of transpiration and extreme sensitivity to moisture deficit (Kriedemann and Barrs, 1981).

Figure 3.9 records a typical daily course of photosynthetic and transpirational leaf activity, on the eastern and western side of two adjacent grapefruit trees. Solar irradiation peaks are on the eastern side during late

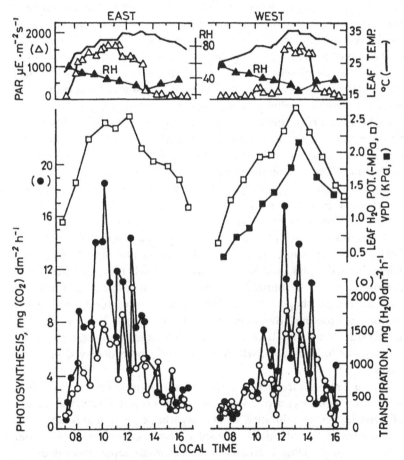

Figure 3.9   Daily course of leaf activity on the eastern (EAST) and western (WEST) side of two adjacent Marsh seedless grapefruit (*C. paradisi*) trees, recorded on September 19, 1979. PAR, photosynthetically active radiation; RH, relative humidity; VPD, vapor pressure deficit; POT, potential. Redrawn from Fishler (1985)

morning and on the western side in the early afternoon. Correspondingly, the highest photosynthetic rates are reached in the eastern side between 0900 and 1200 and in the west between 1200 and 1400. Morning activity rates are generally somewhat higher than afternoon rates (Fishler, 1985). Photosynthetic activity of citrus saturates at relatively low light intensities of 600 to 700 PAR (Syvertsen, 1984; Vu and Yelenosky, 1988), which is about 30% of full sunlight. However, in the orchard these PAR levels are attained only for a few hours and only in external layers of the canopy. In shaded portions of the canopy of certain cultivars (e.g. Marsh grapefruit) light intensities may be less than 1% of full sunlight (Monselise, 1951). The spectral composition of the light filtered through several leaf layers is also different from that of original sunlight. This variability in exposure of the leaves to light makes it difficult to provide reliable, whole-tree irradiation estimates. The daily course of transpiration (Figure 3.9) does not seem to be closely linked to solar irradiation but rather to leaf temperature and leaf water potential. Figure 3.9 shows the occurrence of cyclic oscillations in photosynthesis and transpiration. Such oscillations have been recorded initially under controlled growth conditions by Kriedemann (1968) and then under field conditions by Levy and Kaufmann (1976) but their physiological significance is still not properly understood. These oscillations present another serious problem of attempts to integrate the photosynthetic activity over time, at the single leaf as well as at the whole-tree level.

Although citrus trees thrive in hot, dry environments, leaf photosynthesis has a relatively low temperature optimum (Kriedemann, 1968). Temperatures of 25 to 30 °C are optimal, temperatures of 35 °C and above definitely reduce photosynthetic activity. Extremely high light intensities, as those occurring in subtropical desert areas, cause leaf temperatures to rise considerably beyond the ambient temperature, due to insufficient evaporative cooling (Syvertsen and Lloyd, 1994). This rise in leaf temperature may also be involved in high-irradiation damage to the photosynthetic apparatus, which has recently been observed in citrus leaves.

Moisture stress is frequently encountered in natural habitats as well as in commercial citrus groves. Soil water deficits result in stomatal closure followed by severe inhibition of photosynthesis, presumably involving impairment of the ribulose bisphosphate carboxylase system (Vu and Yelenoksi, 1988, 1989). Increases in the leaf-to-air vapor pressure deficit (VPD), as observed under low relative humidities, are closely correlated with decreases in stomatal aperture and reduction of photosynthesis (Khairi and Hall, 1976; Kriedemann and Barrs, 1981). The sensitivity of the stomatal apparatus to water deficits may represent an evolutionary adaptation towards water conservation (Syvertsen and Lloyd, 1994). Midday depression of transpiration and photosynthesis has been observed during hot summer days in the orchard (Oppenheimer and Mendel, 1934; Sinclair and Allen, 1982). The midday stomatal closure might be associated with an upsurge in VPD but perhaps also with transient leaf water deficits, resulting from difficulties in supplying sufficient soil water to satisfy the high evaporative demands. Under prolonged, severe water stress stomata open in the morning for a short while and then close for the rest of the day (Mendel, 1951).

As already mentioned, integration of the photosynthetic activity of a whole tree is a complex assignment. A first step in this direction has been the assessment of the photosynthetic area of trees by Turrell (1961). In his pioneering study Turrell determined the total leaf area and crown surface area for Valencia orange trees of different ages. In trees of 6 years and older the crown surface represented about 32% of the total leaf area, indicating considerable light interception. When transformed into the current leaf area index (LAI), Turrell's data suggest LAI of 8–15 for Valencia orange. Values of 4.5–11.8 were obtained for other cultivars by Jahn (1979) and Cohen (1984). These values are much higher than those reported for deciduous fruit trees (Jackson, 1980). Integration of whole-tree photosynthetic activity may be obtained by modern computer simulation modeling. The construction of such models takes into account light-saturated photosynthetic rates and seasonal canopy light interception as well as solar radiation and temperature inputs (De Jong and Grossman, 1994). Early models of citrus' photosynthesis have been discussed by Syvertsen and Lloyd (1994).

The export of photosynthate from source leaves of citrus is extremely slow as compared with herbaceous species, and so is the phloem translocation rate. This has been repeatedly shown by $^{14}CO_2$ labeling techniques (Kriedemann, 1969; Wallerstein et al., 1978). In addition to their assimilatory role, citrus leaves function as storage organs. The principal storage carbohydrate of citrus is starch but considerable amounts of soluble sugar are also present. Large amounts of starch reserves are found in all tree organs; leaf starch content may get to 12% dry weight (Goldschmidt and Golomb, 1982). Whereas leaves of annuals show a clear diurnal cycle with all starch removed during night, citrus leaves retain most of their starch and only slight diurnal fluctuations are evident (Goldschmidt et al., 1991). An extensive discussion of citrus' source–sink relations and carbohydrate economy may be found in Goldschmidt and Koch (1996).

The root system

The root system is the hidden part of the plant. Its importance as an anchor in the ground and as a source of water, mineral nutrients and hormones is well recognized. In woody plants roots also serve as major carbohydrate storage organs (Loescher *et al.*, 1990). Differences between root systems assume special significance in stock grafted fruit trees, including citrus.

Roots in general and tree root systems in particular are less accessible to observation and experimentation than aerial, above-ground organs. Many questions regarding root development and physiology remain, therefore, unanswered. It is particularly difficult to follow the behavior of roots in their natural, soil environment. Plexiglas-walled root observation chambers filled with a reconstituted soil profile were used by Bevington and Castle (1985) for detailed observation of citrus root development. The recently introduced use of minirhizotrons (Eissenstat and Duncan, 1992) should facilitate further, advanced study of citrus root systems.

Root morphology and distribution

Citrus plants are taprooted, like most other dicots. During germination the radicle appears first and rapidly grows downward to form a strong, well-defined taproot (Figure 3.2). The taproot is easily recognized at the seedling stage but its identity is often lost in the course of nursery practices and transfer of trees to the orchard.

The structure of the root system of older trees varies greatly with soil, rootstock, irrigation methods and fertilizer supply. According to Castle (1980) typical root systems of citrus acquire a bimorphic nature. Not far from the soil surface one finds a network of strong, lateral roots which provide the supporting framework for a dense mat of fibrous roots. A second, deeper layer of smaller laterals and fibrous roots emerges from the crown in a more-or-less vertical direction. This structure of the root system seems to represent an adaptive strategy. The relatively shallow mass of fibrous roots rapidly absorbs water and applied nutrients from upper soil layers. The second layer of deeper roots is a reserve that prevents extreme drought stress and takes up nutrients not absorbed by the upper root layer. In impervious soils or in those with a high clay content the root system is shallow, with most of the roots concentrated close to the soil surface.

The introduction of drip irrigation and fertilization techniques results in dramatic changes in the structure of root systems. Citrus root systems rapidly adapt to the new conditions, forming a dense core of active roots beneath the dripper. The situation is similar, in a way, to that of container-grown trees, which may be subject to 'root restriction' influences (Bravdo *et al.*, 1992).

Data on citrus root mortality and turnover are scarce due to the difficulty of direct observations. Using minirhizotrons, considerable fibrous root mortality can be demonstrated in mature orchard trees.

Root growth and function

Young, actively growing roots are immediately identified by their white color (Figure 3.10). As roots grow older they acquire a yellow–light-brownish appearance. It has been believed for a long time that only white, actively growing rootlets are involved in uptake of water and mineral nutrients. This view has been disputed, however, and recent evidence suggests that brown roots also participate in uptake processes. Thin, dark-brown fibrous roots are usually dead. The citrus root cap has a typical, pointed shape (Figure 3.11). The occurrence of root hairs in citrus has been debated for a long time (Castle, 1980) but has eventually been convincingly demonstrated by Castle and Krezdorn (1979). The abundance of root hairs and their physiological activity nonetheless need further clarification.

Root growth takes place uninterruptedly, as long as soil temperature, moisture and aeration are adequate, but its intensity varies considerably. Citrus root growth commences in spring, when soil temperatures rise above the biological zero of 13 °C. Root growth resumes earlier in sandy soils that warm up quickly than in heavy, clay soils. Root growth is rather limited at temperatures below 18 °C. Elongation of both pioneer and fibrous roots is linearly correlated with soil temperatures in the range of 18 to 28 °C (Bevington and Castle, 1985). Under orchard conditions most intense root growth is often associated with temperatures of 29 °C and higher (Monselise, 1947). Temperatures above 36 °C seem to restrict root growth (Castle, 1980).

It has long been surmised that periods of intense root growth alternate with flushes of shoot growth (Cossman, 1940; Monselise, 1947; Reed and MacDougal, 1937). This has been clearly shown by Bevington and Castle (1985) (Figure 3.12). Intensive root growth was revealed primarily through emergence of large numbers of lateral, fibrous roots but root elongation rates also increased and means of 6.5 mm per day were recorded.

Citrus root growth is extremely sensitive to soil moisture deficits. Following the withholding of irrigation, root growth stops as soon as the soil matric potential drops to −0.05 MPa. Upon rewatering, recovery of

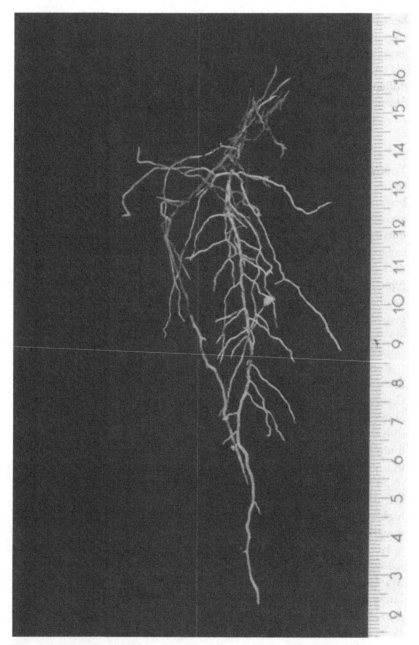

Figure 3.10 Branched lateral roots of calamondin (*C. madurensis*) with white, actively growing rootlets

root growth takes place. However, in previously dry zones of the soil profile there is a two-day lag before the emergence of new rootlets can be observed (Bevington and Castle, 1985). Thus, periodical orchard irrigation regimes produce corresponding root growth cycles.

Fruiting also has marked effects on root development. In trees with a heavy crop load (e.g. during the 'on' year of alternate-bearing cultivars) root growth is completely checked, presumably due to depletion of carbohydrates (Jones *et al.*, 1975; Smith, 1976; Goldschmidt and Golomb, 1982). Root growth of other plant species is also closely dependent upon carbohydrate availability (Farrar and Williams, 1994).

The alternating cycle of root and shoot growth periods (Figure 3.12) suggests the existence of competition for nutrients between roots and shoots. Shoots seem to have the priority; this is also indicated in pruning experiments which show that shoot regrowth occurs at the expense of root growth (Syvertsen, 1994). However, the antagonistic root–top relationship might also have a hormonal background. According to classical concepts, root growth is inhibited by the basipetal stream of auxin produced by growing shoot meristems (Monselise, 1947).

Other plant hormones are also involved in root–top relationships,

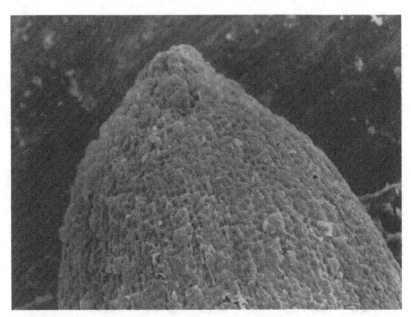

Figure 3.11   Root tip of an actively growing calamondin (*C. madurensis*) root as viewed by SEM

although direct evidence for citrus is generally missing. Young, vigorously growing roots seem to be a major site for the biosynthesis of gibberellins and cytokinins, which are subsequently transported to the top via the xylem (Saidha *et al.*, 1983). Induction of flowering by drought and cold temperatures (see Chapter 4), and the enhancement of fruit degreening by cool temperatures (see Chapter 4) are presumably brought about by reduction in the levels of these root hormones.

Not much is known about the cellular mechanisms of citrus root function. Most of the differences in mineral nutrition and salt tolerance between rootstocks probably reflect distinct structural and physiological properties of their roots. Relatively high hydraulic conductivity of roots can increase the flow of water and mineral elements through the plant system to the leaves. Comparison between citrus rootstocks has shown that rootstocks with high root conductivities generally impart vigorous growth to trees in the orchard. Thus, the relatively vigorous rootstocks, rough lemon and Carrizo citrange, tend to have higher conductivities than the less vigorous Cleopatra mandarin and sour orange rootstocks (Syvertsen, 1981). On the other hand, high conductivity rootstocks appear to be more susceptible to drought, due to the rapid depletion of soil water (Syvertsen and Graham, 1985).

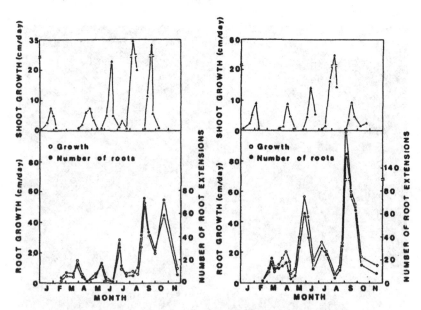

Figure 3.12 Typical pattern of root and shoot growth for Valencia orange trees on rough lemon (*C. jambhiri*) (left) and Carrizo citrange (*P. trifoliata* × *C. sinensis*) (right) rootstocks. Reproduced from Bevington and Castle (1985), with permission

Vesicular–arbuscular mycorrhizal fungi are unique, beneficial fungi which develop symbiotic associations with roots of most plant species, including citrus. Citrus mycorrhizae are widely distributed and every tree is probably infected by mycorrhizal fungi at some stage during its life. Although there are more than 30 species of endomycorrhizal fungi, only a few of these are regularly associated with field grown citrus trees. Best known are several species of *Glomus* (Menge *et al.*, 1977), which are also used as inoculum in controlled experiments. The beneficial nature of the mycorrhizal association with citrus became evident when the addition of mycorrhizal fungi to fumigated or sterilized soil overcame the stunting of certain rootstocks, bringing about a remarkable stimulation of growth.

The beneficial effects of the mycorrhizae result primarily from improved uptake of phosphorus (P) and possibly other microelements. Indeed, the fertilization of poor or sterilized soils with P mimics the mycorrhizal effects to some extent. Mycorrhizal inoculation causes marked increases in cytokinins, which may also account for the enhanced growth effects (Edriss *et al.*, 1984). Citrus rootstocks vary widely in their 'mycorrhizal dependency'. Fast-growing rootstocks reveal greater mycorrhizal dependency than slow ones (Graham and Syvertsen, 1985). Attempts to correlate mycorrhizal association with root structure and additional physiological processes have been only partly successful so far.

Eco-physiological perspective

The genus *Citrus* is believed to have originated in tropical and semitropical parts of South East Asia and spread from there to other continents (see Chapter 1). Different ideas have been proposed, however, with regard to climatic adaptation of citrus. Kriedemann and Barrs (1981) hypothesized that progenitors of present-day cultivars evolved as understory trees in tropical rain forests. This hypothesis is in accordance with the view that citrus is essentially a mesophyte, as indicated by its luxurious foliage and relatively shallow root system.

On the other hand, as also noted by Kriedemann and Barrs (1981), the mature, leathery leaves with their thick cuticle and epicuticular wax reveal considerable xeromorphic adaptation, which is particularly striking in sun leaves. Moreover, citrus transpiration rates are low and the trees can survive rather extended periods of drought. Yet, upon exposure to direct, high irradiance leaf temperatures may rise 8–10 °C above air temperature, due to insufficient evaporative cooling (Syvertsen and Lloyd, 1994). This again indicates that citrus leaves are better adjusted to shaded environments, as also suggested by their high chlorophyll content (Syvertsen and Lloyd, 1994).

The apparently contradictory mesophytic and xerophytic features of citrus may, perhaps, be accounted for by some modification of the 'tropical rain forest origin' hypothesis, as recently proposed by Goldschmidt and Koch (1996). In semitropical as well as in most tropical climate zones there are periods of drought. Whereas major vegetative and reproductive growth activity takes place during periods of abundant moisture, plants are subsequently exposed to shorter or longer periods of drought and must, therefore, have the ability to cope with water-stress situations. The mature, firm textured leaves (but, not the younger, juicy leaves, which are extremely drought sensitive!) seem indeed to be well adapted, both structurally and physiologically, for such harsh conditions. The fruit too, although juicy on the inside, is efficiently protected against water loss by the thick, wax coated peel as well as by the unique structural features of the juice sacs (see Chapter 4). We may thus conclude, that the need of citrus to adjust to periods of drought has resulted in a high priority for water conservation, as revealed by the numerous structural and physiological traits mentioned above.

Recommended reading

Castle, W. S. (1980). *Citrus* root systems: their structure, function, growth and relationship to tree performance. In *Proc. Int. Soc. Citriculture, 1978*, ed. P. R. Cary, pp. 62–9. Griffith, NSW, Australia.

Erickson, L. C. (1968). The general physiology of *Citrus*. In *The Citrus Industry*, Vol. 2, ed. W. Reuther, L. D. Batchelor and H. J. Webber, pp. 86–126. Berkeley: University of California Press.

Goldschmidt, E. E. and Koch, K. E. (1996) Citrus. In *Photoassimilate Distribution in Plants and Crops: Source–Sink Relationships*, ed. E. Zamski and A. A. Schaffer, New York: Marcel Dekker (in press).

Kriedemann, P. E. and Barrs, H. D. (1981). *Citrus* orchards. In *Water Deficits and Plant Growth*, Vol. 6, ed. T. T. Kozlowski, pp. 325–417. New York: Academic Press.

Schneider, H. (1968). The anatomy of citrus. In *The Citrus Industry*, Vol. 2, ed. W. Reuther, L. D. Batchelor and H. J. Webber, pp. 1–85. Berkeley: University of California Press.

Syvertsen, J. P. and Lloyd, J. (1994). *Citrus*. In *Handbook of Environmental Physiology of Fruit Crops*, Vol. 2, ed. B. Schaffer and P. C. Andersen, pp. 65–99. Boca Raton, FL: CRC Press.

Literature cited

Bevington, K. B. and Castle, W. S. (1985). Annual root growth pattern of young citrus trees in relation to shoot growth, soil temperature and soil water content. *J. Am. Soc. Hort. Sci.*, **110**: 840–5.

Bravdo, B. A., Levin, I. and Assaf, R. (1992). Control of root size and root environment of fruit trees for optimal fruit production. *J. Plant Nutr.*, **15**: 699–712.

Castle, W. S. (1980). *Citrus* root systems: their structure, function, growth and relationship to tree performance. In *Proc. Int. Soc. Citriculture*, ed. P. R. Cary, pp. 62–9. Griffith, NSW, Australia.

Castle, W. S. and Krezdorn, A. H. (1979). Anatomy and morphology of field-sampled citrus fibrous roots as influenced by sampling depth and rootstock. *HortScience*, **14**: 603–5.

Cohen, S. (1984). Light relations of an orange canopy. PhD Thesis, The Hebrew University of Jerusalem, 117 pp.

Cossman, K. F. (1940). *Citrus* roots: their anatomy, osmotic pressure and periodicity of growth. *Palest. J. Bot.*, **3**: 65–103.

Davies, F. S. and Albrigo, L. G. (1994). *Citrus*. Wallingford, UK: CAB International, 254 pp.

De Jong, T. M. and Grossman, Y. L. (1994). A supply and demand approach to modeling annual reproductive and vegetative growth of deciduous fruit trees. *HortScience*, **29**: 1435–42.

Edriss, M. H., Davis, R. M. and Burger, D. W. (1984). Influence of mycorrhizal fungi on cytokinin production in sour orange. *J. Am. Soc. Hort. Sci.* **109**: 587–90.

Eissenstadt, D. M. and Duncan, L. W. (1992). Root growth and carbohydrate responses in bearing citrus trees following partial canopy removal. *Tree Physiol.*, **10**: 245–57.

Erickson, L. C. and Brannaman, B. L. (1960). Abscission of reproductive structures and leaves of orange trees. *Proc. Am. Soc. Hort. Sci.*, **75**: 222–9.

Farrar, J. F. and Williams, J. H. H. (1994). Control of the rate of respiration in roots: compartmentation, demand and the supply of substrate. In *Compartmentation of Metabolism*, ed. M. Emes, pp. 167–88. London: Butterworths.

Fishler, M. (1985). Fruit size as related to photosynthesis and partition of assimilates in grapefruit trees. PhD Thesis, The Hebrew University of Jerusalem, 121 pp.

Frost, H. B. and Soost, R. K. (1968). Seed reproduction: development of gametes and embryos. In *The Citrus Industry*, Vol. II, ed. W. Reuther, L. D. Batchelor and H. J. Webber, pp. 290–324. Berkeley: University of California Press.

Goldschmidt, E. E. and Golomb, A. (1982). The carbohydrate balance of alternate-bearing citrus trees and the significance of reserves for flowering and fruiting. *J. Am. Soc. Hort. Sci.*, **107**: 206–8.

Goldschmidt, E. E. and Koch, K. E. (1996). Citrus. In *Photoassimilate Distribution in Plants and Crops: Source–Sink Relationships*, ed. E. Zamski and A. A. Schaffer, New York: Marcel Dekker (in press).

Goldschmidt, E. E., Golomb, A. and Galili, D. (1991). The carbohydrate balance of *Citrus* source leaves: effects of crop load, girdling and diurnal fluctuations. *Alon Hanotea* (Hebrew), **46**: 261–6.

Goren, R. (1993). Anatomical, physiological and hormonal aspects of abscission in citrus. *Hort. Rev.*, **15**: 145–81.

Graham, J. H. and Syvertsen, J. P. (1985). Host determinants of mycorrhizal dependency of citrus rootstock seedlings. *New Phytol.*, **101**: 667–76.

Jackson, J. E. (1980). Light interception and utilization by orchard systems. *Hort. Rev.*, **2**: 208–67.

Jahn, O. L. (1979). Penetration of photosynthetically active radiation as a measurement of canopy density of citrus trees. *J. Am. Soc. Hort. Sci.*, **104**: 557–60.

Jones, W. W., Embleton, T. W. and Coggins, C. W., Jr (1975). Starch content of roots of 'Kinnow' mandarin trees bearing fruit in alternate years. *HortScience*, **10**: 514.

Khairi, M. A. and Hall, A. E. (1976). Temperature and humidity effects on net photosynthesis and transpiration on citrus. *Physiol. Plant.*, **36**: 29–34.

Kozlowski, T. (1964). Shoot growth in woody plants. *Bot. Rev.*, **30**: 335–92.

Kriedemann, P. E. (1968). Some photosynthetic characteristics of citrus leaves. *Austr. J. Biol. Sci.*, **21**: 895–905.

Kriedemann, P. E. (1969). ^{14}C distribution in lemon plants. *J. Hort. Sci.*, **44**: 273–9.

Kriedemann, P. E. (1971). Crop Energetics and horticulture. *HortScience*, **6**: 432–8.

Kriedemann, P. E. and Barrs, H. D. (1981). *Citrus* orchards. In *Water Deficits and Plant Growth*, Vol. 6, ed. T. T. Kozlowski, pp. 325–417. New York: Academic Press.

Levy, Y. and Kaufmann, M. R. (1976). Cycling of leaf conductance in citrus exposed to natural and controlled environments. *Can. J. Bot.*, **54**: 2215–18.

Loescher, W. H., McCamant, T. and Keller, J. D. (1990). Carbohydrate reserves, translocation and storage in woody plant roots. *HortScience*, **25**: 274–81.

Mendel, K. (1951). Orange leaf transpiration under orchard conditions. III. Prolonged soil drought and the influence of stocks. *Palest. J. Bot.*, **8**: 45–53.

Menge, J. A., Nemec, S., Davis, R. M. and Minassian, V. (1977). Mycorrhizal fungi associated with citrus and their possible interactions with pathogens. In *Proc. Int. Soc. Citriculture, 1977*, Vol. 3, ed. W. Grierson, pp. 872–6. Lake Alfred, Florida: ISC.

Monselise, S. P. (1947). The growth of citrus roots and shoots under different cultural conditions. *Palest. J. Bot.*, **6**: 43–54.

Monselise, S. P. (1951). Light distribution in *Citrus* trees. *Bull. Res. Coun. Israel*, **1**: 36–53.

Oppenheimer, J. D. and Mendel, K. (1934). Some experiments on water relations of citrus trees. *Hadar*, **7**: 150–3.

Reed, H. S. and MacDougal, D. T. (1937). Periodicity in the growth of the orange tree. *Growth*, **1**: 371–3.

Saidha, T., Goldschmidt, E. E. and Monselise, S. P. (1983). Endogenous growth regulators in tracheal sap of citrus. *HortScience*, **18**: 231–2.

Schneider, H. (1968). The anatomy of citrus. In *The Citrus Industry*, Vol. II, ed. W. Reuther, L. D. Batchelor and H. J. Webber, Berkeley: University of California Press.

Schroeder, C. A. (1953). Spirality in *Citrus*. *Bot. Gaz.*, **114**: 350–2.

Sinclair, T. R. and Allen, L. H. Jr (1982). Carbon dioxide and water vapour exchange of leaves on field-grown citrus trees. *J. Expt. Bot.* **33**: 1166–75.

Smith, P. (1976). Collapse of 'Murcott' tangerine trees. *J. Am. Soc. Hort. Sci.*, **101**: 23–5.

Syvertsen, J. P. (1981). Hydraulic conductivity of four commercial citrus rootstocks. *J. Am. Soc. Hort. Sci.*, **106**: 378–81.

Syvertsen, J. P. (1984). Light acclimation in citrus leaves. II. Assimilation and light, water and nitrogen efficiency. *J. Am. Soc. Hort. Sci.*, **109**: 812–17.

Syvertsen, J. P. (1985). Hydraulic conductivity of roots, mineral nutrition and leaf gas exchange of citrus rootstocks. *J. Am. Soc. Hort. Sci.*, **110**: 865–9.

Syvertsen, J. P. (1994). Partial shoot removal increases net CO_2 assimilation and alters water relations of *Citrus* seedlings. *Tree Physiol.*, **14**: 497–508.

Syvertsen, J. P. and Graham, J. H. (1985). Hydraulic conductivity of roots, mineral nutrition and leaf gas exchange of citrus rootstocks. *J. Am. Soc. Hort. Sci.*, **110**: 865–9.

Syvertsen, J. P. and Lloyd, J. (1994). *Citrus*. In *Handbook of Environmental Physiology of Fruit Crops*, Vol. 2, ed. B. Schaffer and P. C. Andersen, pp. 65–99. Boca Raton, FL: CRC Press.

Turrell, F. M. (1961). Growth of the photosynthetic area of citrus. *Bot. Gaz.*, **122**: 285–98.

Turrell, F. M., Young, R. H., Austin, S. W. and Garber, M. J. (1980). Growth of woody frame of the grapefruit tree (*Citrus paradisi* Macf.). In *Proc. Int. Soc. Citriculture, 1978*, ed. P. R. Cary, pp. 325–82. Griffith, NSW, Australia.

Vu, J. C. V. and Yelenosky, G. (1988). Solar irradiance and drought stress effects on the activity and concentration of ribulose biphosphate carboxylase in 'Valencia' orange leaves. *Isr. J. Bot.*, **37**: 245–56.

Vu, J. C. V. and Yelenosky, G. (1989). Non-structural carbohydrate concentrations in leaves of 'Valencia' orange subjected to water deficits. *Environ. Expt. Bot.*, **29**: 149–54.

Wallerstein, I., Goren, R. and Monselise, S. P. (1978). Rapid and slow translocation of ^{14}C-sucrose and ^{14}C-assimilates in *Citrus* and *Phaseolus* with special reference to ringing effect. *J. Hort. Sci.*, **53**: 203–8.

4

Reproductive physiology: flowering and fruiting

The flowering of citrus

Introduction

THE TRANSITION OF the vegetative, leaf producing meristem into the reproductive floral meristem is the initial event in the long chain of developmental processes leading to seed and fruit production.

The environmental and endogenous control of flower bud differentiation is quite complex and varies considerably from one species to another. Citrus trees, like other fruit trees, are polycarpic plants undergoing repeated cycles of flowering and fruiting. Fruit trees never commit all their buds to flowering – a certain number of buds must be retained in the vegetative, non-differentiated state to ensure the tree's future. This raises certain questions with regard to the nature of floral induction in fruit trees. The suggestion has been made that flowering in fruit trees might be under 'negative' control, i.e. all buds are induced to flower but their actual flowering is controlled by the presence of a 'flowering inhibitor'. This hypothesis must be weighed against the more common concept of a positive 'floral stimulus' which must reach the apex to start the process of flower bud differentiation (Lang, 1965).

Certain aspects of citrus' floral development derive from the nature of citrus as a tropical–subtropical evergreen which, unlike deciduous fruit trees, does not have true dormancy (Monselise, 1985). Deciduous fruit trees form flower buds during early summer. These flower buds complete their morphological development prior to the onset of winter dormancy and appear to be ready by the fall for the burst and bloom of the following spring. In citrus, flower bud differentiation starts during the winter and moves without interruption towards floral development and bloom. One notable exception is *Poncirus trifoliata*, a monospecific genus belonging to the Citrinae. This species has scale-protected flower buds which are initiated during summer. Bloom occurs in early spring, on leafless branches which have completely shed their characteristic trifoliate leaves at the beginning of winter. Thus, *Poncirus* has a consistent deciduous habit, probably caused by its adaptation to the cooler climates of the temperate zones of north-eastern Asia (Monselise, 1985).

Research on flowering has focused to a large extent on plants in which flower bud differentiation is induced photoperiodically. In numerous plant species flowering does not appear to be under photoperiodic control, however, and floral induction is triggered by temperature or other environmental factors. In still other plants, flower bud differentiation does not seem to depend entirely upon environmental factors and it is assumed, therefore, to be controlled 'autonomously' by endogenous factors (Halevy, 1984). Elucidation of the role of environmental factors in the control of flowering is possible only through experimentation under controlled environment conditions. As pointed out by Davenport (1990), most research on the flowering of citrus has not been conducted under such conditions. While keeping this reservation in mind we shall proceed now to discuss the evidence concerning the control of flower bud differentiation in citrus.

Environmental control

All species of citrus begin their major flowering flush in subtropical regions during the late winter months, when days are short. The suggestion has been made, therefore, that flowering of citrus might be induced by the short winter day length.

Field experiments and greenhouse observations indicate that short days are insufficient, in themselves, to induce flowering of citrus. It seems, rather, that short photoperiods might predispose the induction of flowering by low temperatures. Lenz (1969) described experiments in which rooted cuttings of Washington navel orange were exposed to 8, 12 or 16 hour photoperiods at either mild (24 °C day/19 °C night) or warm (30 °C day/25 °C night) temperature regimes for a continuous period of 34 weeks. Plants growing in the higher temperature regime produced vegetative shoots with no inflorescences, regardless of day length. Plants growing in the mild temperature range produced inflorescences only in 8- and 12-hour day lengths. However, at lower temperatures (18 °C day/13 °C night), flowering could be obtained under both short and long day conditions (Moss, 1977). Vegetative growth and stem elongation, on the other hand, were strongly promoted under long days. While these results substantiate the role of low temperatures, they also point to the existence of some kind of short-day flowering response. The precise critical day length has not been determined and it is difficult to say, therefore, to what extent the photoperiod actually restricts the flowering of citrus under natural and agricultural habitats.

Chilling winter temperatures have been suspected for many years to be involved in the floral induction of citrus. Under most natural conditions, however, cool temperatures prevail concomitantly with short photo-periods, and the specific effects of cool temperature cannot be easily determined. The observations of Cassin *et al.* (1969) have indicated that cool temperatures occurring in high altitudes bring about flowering even in the tropics, where changes in day length are small. Experiments under controlled conditions have shown that cool temperature regimes (18–15 °C/13–8 °C; day/night) induce flowering in rooted cuttings of orange (Moss, 1969; Lovatt *et al.*, 1988) and lime (Southwick and Davenport, 1986). Longer exposure to chilling temperatures (two to eight weeks) increased the total number of shoots as well as the intensity of flowering. Milder temperatures (24/19 °C; day/night) produced fewer, and mostly leafy, inflorescences (Moss, 1969). While most studies indicate that the chilling treatment is perceived by the aerial part of the plant (leaves, twigs, buds), some reports suggest a role for the roots as well (Moss, 1977; Davenport, 1990).

Water stress seems to be the major flower-inducing signal under semi-tropical conditions, possibly representing many of the original, native habitats of citrus is South East Asia. This has been clearly demonstrated by Cassin *et al.* (1969), who analyzed bloom and rainfall data for numerous locations in the tropics. Flowering invariably occurred follow-ing the renewal of rain, subsequent to a shorter or longer period of drought. Water stress has been, and is still being, used in Italy, Israel and California for the production of summer lemons, 'Verdelli' (Casella, 1935). Withdrawal of irrigation for about two months during midsummer is followed by a wave of blooming towards the fall. The resultant fruit develop during winter and are ready for harvest by early summer, when lemons are highly priced. Water stress has been used traditionally to induce flowering in lemons and related cultivars which have a natural tendency to produce some flowers throughout the year. There is no obvious reason, however, why this agrotechnique should not be used for the production of an out-of-season crop in other desirable cultivars, since the basic flowering-control mechanisms seem to be common to all citrus species.

It is clear from the foregoing that two environmental stimuli play a major role in the natural control of citrus flowering. Water stress, mainly under tropical conditions, and chilling temperatures, in the cooler, subtropical growing areas. Southwick and Davenport (1986) compared water-stress and low-temperature treatments, following also the changes in leaf xylem pressure potentials. They were able to show that the flower-inducing, low-temperature treatments did not involve water stress to any extent; in fact, control trees growing in the greenhouse had lower midday pressure potentials than the cold treated trees. Thus, low temperatures and water stress appear to be separate flower-inducing signals. It can still be questioned, however, whether these environmental signals should be regarded as truly inductive mechanisms. It may be argued that some cessation of vegetative growth, which is common to both low temperatures and water stress, is all that is needed for the triggering of flower bud differentiation of citrus.

Nutritional and hormonal control

Carbohydrate levels have been suggested as playing a role in the control of flowering (Goldschmidt et al., 1985). The evidence supporting this notion is correlative and indirect, so far. Whereas depletion of carbohydrate reserves due to heavy fruit load interfered with flower bud differentiation of the subsequent season (Goldschmidt and Golomb, 1982), girdling (removal of a ring of bark from trunk or scaffold branches), known to cause accumulation of carbohydrates above the girdle, markedly increased flowering when performed during autumn (Cohen, 1982; Goldschmidt et al., 1985). On the other hand, carbohydrate levels did not correlate well with the cool-temperature promotion of flowering (Goldschmidt et al., 1985). It is not clear, therefore, whether carbohydrates have a specific regulatory role in flowering or whether, perhaps, only a threshold level of energy-providing carbohydrates is required for flower bud induction.

A link between nitrogen metabolism and flowering has also been indicated in recent years. Induction of flowering by low temperatures or water stress was correlated with an increase in leaf ammonia content. Also, partial induction by moderate water stress could be intensified through foliar application of urea, resulting in elevated ammonia levels. Presumably, the accumulation of ammonia during stress leads to increased biosynthesis of arginine polyamines, which play a role in the meristematic activity involved in flower bud differentiation (Lovatt et al., 1988).

Considerable efforts have been dedicated to the search for a hormonal control system of flowering in citrus. Work in this area is based almost entirely on applications of plant growth substances and growth retardants and the evidence is, therefore, rather indirect. Nevertheless, some major trends emerge quite clearly out of this work.

Of the large number of compounds tried, only gibberellins proved to have a consistent, reproducible effect. Since the early report of Monselise and Halevy (1964) and following numerous detailed studies thereafter (Moss, 1970; Goldschmidt and Monselise, 1972; Lord and Eckard, 1987), gibberellin A_3 (GA_3) as well as other gibberellins have been shown to inhibit the flowering of citrus. Amounts as low as 0.1 nmole of GA_3 or $GA_{4/7}$ per bud caused a 75% reduction in the number of flowers; higher doses prevented flowering altogether. Inhibition of flowering by GA_3 has also been reported in other fruit tree species and woody perennials (Goldschmidt and Monselise, 1972; Pharis and King, 1985).

Assuming that endogenous gibberellins act, like exogenous gib-berellins, as inhibitors of flowering, one may expect that application of gibberellin-biosynthesis inhibitors (growth retardants) will enhance the flowering of citrus. Early work with growth retardants gave variable results but the recently introduced triazole compounds (paclobutrazol) fairly consistently promote the flowering of citrus cultivars, whether applied by spraying or soil applications (Harty and van Staden, 1988; Greenberg et al. 1993).

Among the internal factors which regulate flowering the fruit must be mentioned. The presence of fruit strongly inhibits flower formation and any shoot that bears fruit seldom produces flowers (Moss, 1977). Endogenous, fruit-gibberellins may also be involved in this phenomenon.

Flower bud differentiation in citrus appears to be intimately related to the development of shoot types. Three main shoot types can be dis-tinguished: vegetative shoots, leafy inflorescences and purely generative, leafless inflorescences (Figure 4.1). Stem elongation and the intensity of flowering appear to be inversely correlated: vegetative shoots are longest, leafy inflorescences are intermediate in length and the purely generative, leafless inflorescences are shortest (Figure 4.1). Treatment with low amounts of GA_3 shifts the balance of shoot types towards the vegetative end, bringing about a higher percentage of vegetative shoots (Gold-schmidt and Monselise, 1972). Paclobutrazol, on the other hand, shifts the balance towards the generative end, leading to high percentages of pure, leafless inflorescences (Greenberg et al., 1993).

These observations strongly support the notion that endogenous gib-berellins play a major role in the control of citrus flowering. Gibberellins may be assumed to act as native quantitative inhibitors of flowering, maintaining a delicate balance between vegetative and generative shoot types and inflorescences. Direct evidence concerning the qualitative and quantitative nature of endogenous gibberellins in citrus buds during their early, predifferentiation stage is still missing.

A broad array of circumstantial evidence seems to be in accordance with the proposed role of endogenous gibberellins as inhibitors of flowering. Growing root tips are supposedly major sites of gibberellin biosynthesis. Low temperatures as well as withdrawal of irrigation bring about a cessation of root growth, thereby restricting the supply of gibberellins to the canopy. This should be the mechanism underlying both low temperature and water stress induction of flowering. Conversely, uninterrupted root growth as found in the humid tropics may interfere with flower bud differentiation through year-round supply of high levels of endogenous gibberellins.

Are gibberellins the only hormonal system involved in the regulation of flowering in citrus?

The available evidence does not allow a definite answer to this question, although the control of the developmental events leading to flowering probably also involves other hormones (Davenport, 1990). Auxins may

Figure 4.1   Vegetative (A), leafy inflorescence (B) and leafless inflorescence (C) shoots in lemon (C. limon)

effect flowering through their role in apical dominance, as suggested by the fact that bending of twigs increases flowering in lemon. Further experimental work will be required to clarify the role of other growth substances in the control of flowering in citrus.

From flower bud differentiation to anthesis (bloom)

Flowering shoots are produced in citrus on woody twigs of the previous year's spring flush but may also be borne on younger, summer flush twigs or on older wood. Flower bud differentiation and inflorescence ontogeny have been described in detail by Lord and Eckard (1985) (Figures 4.2, 4.3). The first observable change in the differentiating bud is a flattening of the apical dome in the terminal meristem, but at this stage the meristem is not yet irreversibly committed to flowering. Reversal may be obtained through application of GA_3 as long as sepals have not formed (Lord and Eckard, 1987).

The inflorescences of citrus are cymose; the terminal flower is initiated first and the lateral axillary flowers later. Not much is known about the differentiation of these lateral flower buds. Lord and Eckard (1987) propose that messages coming from the terminal apex act to determine the fate of the axillary, lateral meristems. Early commitment of the shoot apex to flowering leads to totally generative, leafless inflorescences (type C, Figure 4.1). Leafy inflorescences arise from delayed commitment to flowering by the shoot apex which brings about reduced axillary flower development or, under extreme cases, the formation of a single, terminal flower. This hypothesis is supported by field data indicating that bud break and anthesis of leafless inflorescences precede those of leafy inflorescences by at least several days (Lovatt et al., 1987). This difference in time of anthesis might be meaningful with regard to climatic conditions affecting fruit set and subsequent fruit development (Haas, 1949; Moss, 1977).

Among the lateral flowers, the first subterminal is the most retarded compared with the terminal flower. A progressive increase in rates of growth occurs in the primordia of successive nodes distal from the apex. Indeed, the most distal flower is usually the most developed and the first (after the terminal flower) to reach anthesis, whereas the subterminal flower is the latest to bloom (Zacharia, 1951; Lovatt et al., 1987). Figure 4.4 demonstrates these relationships for leafless as well as for leafy inflorescences. This is evidently an apical-dominance phenomenon, well documented also in other cymose inflorescences.

Dates and duration of bloom are highly variable even for the same

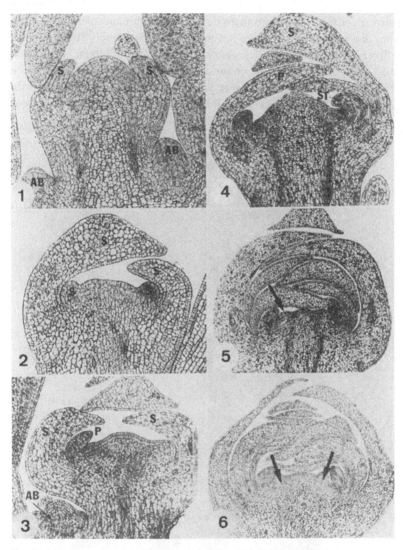

Figure 4.2   Histogenesis in the terminal flower of Washington navel orange (*C. sinensis*). 1. Sepal initiation begins; note bud primordia in axils of the foliar leaves. (×132). 2. Later sepal initiation; note fattening of apex (×84). 3. Petal initiation; note axillary bud differentiation (×53). 4. Stamen initiation; note conical shape of apex (×53). 5, 6. Carpel initiation (arrows) (both ×33). AB, axillary bud; P, petal; S, sepal; ST, stamen. Reproduced from Lord and Eckard (1985), with permission

cultivar – differences of up to 40 days in the commencement of anthesis from one year to the next are not uncommon. Seemingly slight climatic differences between locations also affect the dates of bloom. Within the same tree, the south-west quadrant (in the northern hemisphere) is usually the first to open its flowers, whereas the north-east quadrant is the last. Observations in Israel and California have shown that the rate of

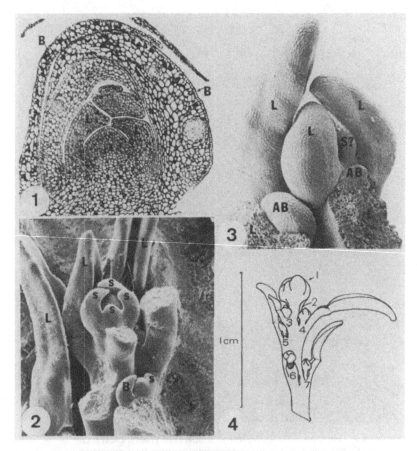

Figure 4.3 Inflorescence development in Washington navel orange (*C. sinensis*). 1. Resting bud with bracts, foliar leaf primordia and promeristem (×63). 2. SEM of breaking inflorescence shoot showing developing buds in axils of foliar leaves and possible sepal primordium (×60). 3. Flowering shoots with initiated sepals on the terminal flower and an axillary bud with initiated bract and sepals being formed on lateral floral apex (×60). 4. Diagram of inflorescence showing the flower positions, numbered basipetally. At this stage the terminal flower would be initiating carpels. AB, axillary bud; B, bract; L, foliar leaf; S, sepal. Reproduced from Lord and Eckard (1985), with permission

Figure 4.4 Leafy and leafless inflorescences of Marsh seedless grapefruit (*C. paradisi*), showing the development of lateral flowers in relation to their distance from the terminal flower

flower development from budbreak to anthesis is rather closely dependent upon the accumulation of heat units above a minimum threshold temperature (Lomas and Burd, 1983; Lovatt *et al.*, 1987).

Floral development and duration of bloom have been modelled for Washington navel orange in California, using a 9.4 °C threshold temperature and January 29 as the starting point for heat unit accumulation (Bellows *et al.*, 1989). The duration of the flowering period is also largely dependent upon the prevailing temperatures. Warmer than usual weather will bring about opening of flowers within a few days, resulting in a concentrated wave of bloom, petal fall and fruit set (Figure 4.5). Cool spring weather, on the other hand, may lead to an extended period of diffuse flowering. Such seasonal differences may have important consequences for the chances of pollination and fruit set, particularly in self-incompatible cultivars (e.g. mandarin hybrids) where overlapping with pollination is critical.

Floral morphology and biology

The general structure of a citrus flower at anthesis is outlined in Figure 4.6. Flowers ready to open are 1.5–3.0 cm long, supported by a pedicel. The calyx is cup-like with five sepals. In the immature, ball-shaped flower bud the distal ends of the sepals envelop the internal floral organs. The sepals are pushed aside by the expanding corolla. The corolla has five white petals alternating with the sepals. Petals are thick, glossy with interlocking, marginal, papillae to keep them reflexed. The stamens appear as 20–40 white, partially united filaments, each bearing a yellow, four-lobed anther. Anthers surround the pistil at or close to the level of the stigma. The floral disc secretes watery nectar through the stomata. The ovary (eight to 14 carpels), style and stigma comprise the pistil. Carpel development has been described by Ford (1942) and Schneider (1968). At

Figure 4.5 Three single flower, leafless inflorescences of Palestine sweet lime (*C. limettoides*) at different stages of bloom. Right, an elongated flower, about to open. Left, an open flower. Middle, the ovary and style after petal fall; this is the early fruit set stage

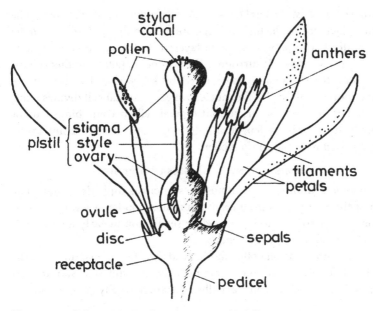

Figure 4.6 Schematic drawing of an open citrus flower

the inner angle of the locule of each carpel the placenta develops, bearing the ovules (see Figure 4.7).

The ovule is anatropous, with the micropyle facing the axis of the ovary. The mature ovule consists of the funiculus, the nucellus, an eight-nucleate embryo sac and two integuments. At flowering the ovary is subglobose, distinct from a narrow style, as in orange, or subcylindrical, merging into the style, as in lemon. The cylindrical style expands into the stigma. The relatively large stigma is receptive one to a few days before anthesis and in some cases up to six to eight days after anthesis (Randhawa *et al.*, 1961). Modified epidermal cells on the stigma secrete a viscous fluid, to which the pollen stick. Canals extend from each locule through the style opening on the stigma surface. Germinated pollen tubes supposedly pass down the canals to convey the two sperm nuclei, although certain reports argue that pollen tubes grow through the spaces between the stylar canals (Banerji, 1954; Geraci *et al.*, 1980).

Micro- and megasporogenesis have been described in detail by Frost and Soost (1968).

ARCHESPORIUM AND POLLEN MOTHER CELLS

At an early stage in the development of the anther, archesporial cells are recognizable by their larger nuclei and different staining. They divide periclinically to produce an outer layer of parietal cells and an inner layer of sporogenous cells. Parietal cells divide to produce four cell layers. The innermost layer forms the tapetum, the others with the epidermis form the anther wall. The tapetum is single layered, but may be two or three layered (Banerji, 1954). It surrounds the cylinder of pollen mother cells or microsporocytes (Osawa, 1912) formed by successive divisions of the primary sporogenous cells. The nucleus of each tapetal cell divides once; later, further divisions occur, so that tapetal cells become binucleate or multinucleate. While pollen grains develop, tapetal cells disintegrate, supplying food material to the pollen.

MEIOSIS AND MICROSPORES IN DIPLOIDS

The monoploid number of chromosomes in citrus diploids is nine. The length of the small chromosomes at maximum condensation in the first division is 2 microns or less. Just before separation in early anaphase I, they may stretch to a greater length.

Before the first division pollen mother cells are distinguishable from the tapetal cells by their size, the single nucleus, and different staining. At anaphase I, the nine bivalents usually separate normally resulting in two

Figure 4.7   A longitudinal section through an ovary of lemon (*C. limon*) during anthesis, as viewed by SEM (×30). The locules are empty except for the ovules, which are attached to the central axis. Oil glands appear along the external layer of the peel

daughter nuclei. Secondary association of the chromosomes during the beginning of the first metaphase has been reported, including description of multivalents, inversion and univalents. Iwamasa (1966) has described translocation in Valencia, inversion in Mexican lime and asynapsis in Mukaku Yuzu.

At the end of the second division, the four sets of nine chromosomes each organize four nuclei within the rounded wall of the mother cell, producing the four-nucleate state. Walls form between the four nuclei, and the four new cells separate and round off. This is the microspore tetrad stage.

POLLEN GRAIN AND SPERM CELLS (MALE GAMETES)

Development follows the usual course for angiosperm pollen. The microspore enlarges, developing two heavy coats, exine and intine. Before the anther dehisces, the nucleus divides forming a vegetative or tube nucleus and a generative nucleus. The pollen are binucleate, but Banerji (1954) reports grapefruit pollen to be trinucleate. Normal anthers are bright yellow when mature. Defective pollen has a lighter color. Anthers containing no pollen are pale cream and do not dehisce.

Each of the four lobes of the anther develops a microsporangium or enclosure in which microspores form and develop into pollen grains. The two microsporangia in each half of the anther coalesce. In the mature anther the pollen is held in two pollen sacs or anther locules (microsporangia of Satsuma often remain separate). Each half of the anther dehisces by a longitudinal split between the lobes; the epidermis dries out and rolls back the anther wall, exposing the enclosed pollen.

ARCHESPORIUM AND EMBRYO SAC MOTHER CELLS

One cell near the apex of the nucleus is distinguished by its greater size and larger nucleus. This is the archesporial cell. It divides once. The outer cell is the tapetal cell, the inner is the embryo-sac mother cell, or megasporocyte. The embryo-sac mother cell is buried in tissue near the center of the nucellus. Occasionally, more than one embryo-sac mother cell is formed in an ovule.

MEIOSIS AND MEGASPORES

The embryo-sac mother cell grows to several times its original size and becomes elongated. Chromosomes of its nucleus pair during the prophase of the first division. This occurs before the ovule is fully developed (Bacchi, 1943). After the second division, cell walls are formed, producing four cells in a row extending longitudinally in the nucellus. These four cells are the megaspores.

EMBRYO SAC AND EGG CELL

Only the lowermost megaspore develops. It grows longitudinally and develops further to produce the embryo sac. The cytoplasm does not increase and large vacuoles appear. As the megaspore grows, the nucleus divides. The two daughter nuclei go to opposite ends of the embryo sac.

Each nucleus divides again twice. The embryo sac now contains four nuclei near each end. Three remain near the basal (chalazal) end, where they organize the three antipodal cells in the micropylar end – one is the egg cell, two are synergids. At this stage the egg is mature and ready for fertilization. The two remaining nuclei, one near each end of the embryo sac, are the polar nuclei. They move to the middle of the sac, and fuse to form the endosperm nucleus.

The corresponding stages of development are attained later in the ovary than in the anthers of a given flower; microspores begin to develop into pollen grains before the megasporocyte has passed the prophase of the first division (Osawa, 1912).

At the time of opening of the flower, the embryo sac is at the eight-nucleate stage, but may occasionally be in a less advanced stage (Bacchi, 1943).

Opening of flowers begins with partial separation of the tip of one petal. The opening of citrus petals has been shown to be a growth reaction controlled by auxin. At about the same time that flower opening commences the anthers begin to dehisce, so that flowers have to be emasculated prior to opening for controlled pollination. Stigma receptivity, the so-called 'milk drop' stage, lasts for about 3 days. From the day of anthesis, the countdown begins, which will end for each single flower either with abscission or with initial fruit set. The decision is reached about one week after anthesis – by that time petals and anthers would already have withered and abscised (Figure 4.5). In persisting flowers, the appearance of a brown ring between the ovary and the style is the first sign of fruit set. Style abscission, which takes place 7 to 10 days later, is the last event in the flower-to-fruit transition. Style abscission does not occur in certain varieties of citron (*C. medica*) and bergamot (*C. bergamia*) which retain the style throughout fruit development and maturation (Figure 4.8). The persistent style has been regarded as an important trait of the citron for its use in the ritual of the Jewish feast of Tabernacles.

Pollination and fertilization

Pollination consists of pollen transfer to the stigma. The pollen tube germinates, and penetrates the embryo sac in the ovule. Citrus pollen are sticky and adherent, as in other insect-pollinated plants. Citrus flowers are attractive to insects due to abundant pollen, nectar, typical perfume, and the conspicuous corolla. Most citrus species are valuable honey-producing plants. While thrips and mites also abound on flowers, honey bees are the main agent in natural cross pollination. Wind is a minor factor in citrus pollination. Self pollination may occur in self-compatible genotypes by wind-blown pollen or by direct contact of anthers with stigma (more often in protandrous cultivars). Cases of self incompatibility are discussed in Chapter 6. Temperature has considerable effect on pollination efficiency, affecting the rate of pollen-tube growth as well as bee activity. Pollen viability and ovule fertility are also influenced by temperature.

Fertilization (fecundation) is attained by fusion of a sperm (pollen) nucleus with an egg nucleus. Two microgametes are produced by the generative nucleus of the pollen. One microgamete fuses with the egg nucleus producing the zygote, while the other fuses with the two polar

Figure 4.8 A mature citron (*C. medica*) fruit with a persistent style (c. ×0.75)

nuclei initiating the endosperm. Fertilization of the egg cell occurs two or three days after pollination under favorable conditions but, in some cases, a three to four week lapse has been reported. Cell division of the zygote starts soon afterwards. By that time the endosperm is already multicellular.

Nucellar embryos

Nucellar embryos develop asexually by ordinary mitotic division of cells of the nucellus. The apomictic process thus generates seeds containing embryos of a purely maternal genetic constitution.

In apomictic citrus genotypes sexual and apomictic processes occur within the same ovule. Nucellar embryos are initiated from the nucellar tissue in the region around the developing sexual embryo sac (Wakana and Uemoto, 1987, 1988; Koltunow, 1993). Nucellar cells destined to become embryos have large nuclei and a dense cytoplasm (Kobayashi et al., 1982; Bruck and Walker, 1985). Nucellar embryo initial cells are first apparent at or soon after anthesis in the nucellar cell layers surrounding the chalazal portion of the sexual embryo sac. Further initials are found, later, in the micropylar region and along the length of the embryo sac. As a result of growth in the chalazal end of the ovule both sexual and nucellar embryos are eventually driven toward the micropylar end of the embryo sac. The first division of the citrus nucellar initial cells has been found to occur around the time of the first zygotic division. Nucellar embryo development is quite similar to stages of development found in sexual embryos (Bruck and Walker, 1985). The growth of the zygotic embryo is often slower when compared with that of nucellar embryos (a crowding effect can also be observed). The zygotic embryo may also not complete its development (Frost and Soost, 1968). Polyembryonic seed often contain embryos at different stages of maturation.

Not all ovules within a particular citrus ovary are fertilized. Wakana and Uemoto (1988) found that the nucellar embryos in unfertilized ovules were arrested at the globular or very early cotyledonary stages. Koltunow (1993) observed embryo development in fertilized and unfertilized ovules of Valencia orange and found that the initiation of nucellar embryo development occurs in the developing fruit at a similar time in both kinds of ovules. Koltunow assumes that there is a general ovary signal for nucellar embryo development which is independent of fertilization. When unfertilized ovules were excised from mature fruit and cultured on a simple medium lacking plant hormones (Moore, 1985) the arrested embryos were able to complete their development. It appears, therefore, that a sufficient source of nutrition is essential for the completion of nucellar embryogenesis. Presumably, the nutritional factors are normally supplied by the endosperm formed in the embryo sac following double fertilization (Koltunow, 1993).

The significance of apomixis (nucellar embryony) in citrus for evolution, propagation, dissemination, maintenance of heterozygosity (see also Cameron and Frost, 1968) and breeding has been pointed out in Chapters 2, 5 and 6.

Polyembryony is a development of two or more embryos in one seed. Extra embryos are commonly produced apomictically from cells of the seed parent (nucellar embryony) and rarely by production of two or more zygotic embryos, by fusion of the egg or from additional functional embryo sacs in the ovule (Bacchi, 1943).

Parthenocarpy

Fertilization leading to seed formation is generally a prerequisite for fruit set and lack of fertilization will inevitably end up in drop of the ovary. There are, nevertheless, numerous plants which produce seedless fruit. Production of fruit without seeds is parthenocarpy (Frost and Soost, 1968). The setting of fruit without any external stimulation is defined as autonomic parthenocarpy. The Satsuma mandarin, which forms fruit in the absence of pollination, belongs in this category. The term stimulative parthenocarpy is used to describe cases in which some kind of stimulus is required. In stimulative parthenocarpy, pollination, pollen germination and pollen tube growth, unaccompanied by fecundation, provide sufficient stimulation for set of seedless fruit. Thus, self pollination may exert a sufficient stimulus in self-incompatible genotypes for the setting of seedless fruit. Application of gibberellin A_3 replaces the need for pollination in Clementine and in other *C. reticulata* hybrids, leading to production of parthenocarpic, seedless fruit (Chapter 5, Table 5.9). In some cases of parthenocarpy, fruit with occasional seeds can be found as a result of incomplete female sterility (Washington navel orange, Marsh seedless grapefruit). Parthenocarpic tendency and ovule sterility may vary independently. Some usually seeded cultivars may be capable of a variable degree of parthenocarpy, especially self-incompatible ones (see Chapter 6). Vardi *et al.* (1988) state that the potential for pollen-stimulated parthenocarpic fruit is rather widespread in citrus, and the possibility of only a few genes being involved cannot be excluded. Ovule fertility and the presence of compatible pollen mask stimulative parthenocarpy. In natural and induced seedlessness (see Chapter 6), the seedless condition is generally accompanied by irregularities of meiosis. In a few cases in *Citrus*, a phenomenon resembling stenospermocarpy (fecundation followed by post-zygotic abortion) has been noted.

For a cultivar incapable of seed production to be horticulturally acceptable, a high parthenocarpic tendency is essential.

Fruit development and maturation

Fruit structure

The fruit of citrus is a special type of berry termed 'hesperidium'. It is a true fruit arising through growth and development of the ovary, consisting of a variable number (three to seven in *Fortunella*, eight and up in *Citrus* and *Poncirus*) of united, radially arranged carpels. Phylogenetically, carpels are considered by most authors to be modified leaves oriented vertically with their margins curved to join the central axis, thereby forming locules (segments) in which seeds and juice sacs develop (Figure 4.9).

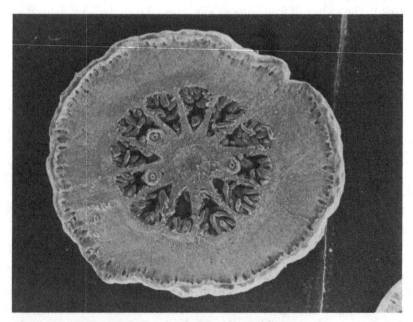

Figure 4.9 Cross section of a Murcott tangor fruitlet (diameter = 10 mm, *c*. 45 days after anthesis), as viewed by SEM (×10). The locules contain seed, attached to the central axis and young juice vesicles, projecting from the sides and from the distal regions of the locular membranes. The peel consists mainly of albedo. Oil glands appear along the external layer of the peel, the flavedo

A small, secondary fruit (navel) is sometimes present at the stylar end of the main fruit. Whereas in certain types of mandarins the navel appears as a tiny, undeveloped fruit, in Washington navel orange the secondary fruit attains a diameter of 2–3 cm, slightly protruding, although still enclosed by the peel of the main fruit (Figure 4.10). The ontogeny of the navel has been described in detail by Lima & Davies (1984).

Citrus fruits are composed of two major, morphologically distinct regions – the *pericarp*, commonly known as the 'peel' or 'rind' and the *endocarp*, which is the edible portion of fruit, often called the 'pulp'. A further distinction is made within the peel; the external, colored portion is the *epicarp*, mostly known as the *flavedo*, whereas the internal, white layer of the peel is the *mesocarp*, generally known as the *albedo*. The 'flavedo' is composed of the cuticle-covered epidermis and a few compactly arranged parenchyma cell layers adjacent to it. Embedded in the flavedo are multicellular, schyzolysogenic oil glands containing essential oils

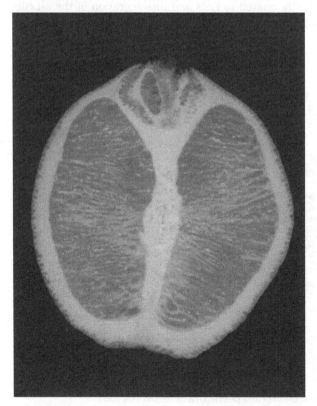

Figure 4.10   A longitudinal section of a mature fruit of Washington navel orange (*C. sinensis*) showing the navel at the stylar end (c. ×0.7)

(Figure 4.11). During early stages of fruit development the flavedo is a dark green, photosynthetically active tissue, with a relatively small number of stomata (20 to 40 mm^{-2}). As the fruit approaches maturation, chlorophyll is gradually lost and chloroplasts are transformed into carotenoid-rich chromoplasts (Goldschmidt, 1988). The deeper layers of the flavedo merge into the white, spongy 'albedo'. In the mature fruit the albedo consists of large, deeply lobed cells with very large intercellular spaces and scattered vascular elements (Figure 4.12). During the early phase of fruit development, when peel growth predominates, the albedo may occupy 60 to 90% of fruit volume. Later, when pulp growth takes over, the albedo becomes thinner and the portion of the albedo declines. In numerous mandarin and orange cultivars the albedo gradually degenerates and disappears, leaving only a net of vascular elements between the flavedo and the pulp – this is the 'reticulum' for which mandarins have been named *C. reticulata*. Physiological diseases such as 'creasing' and 'puffing' are evidently related to lysis and disintegration of the albedo, but the underlying biochemical processes are still poorly understood.

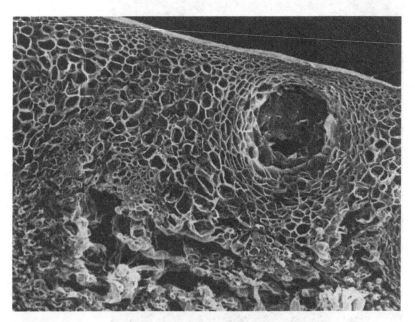

Figure 4.11   Cross section of fruit peel from a senescent Murcott tangor fruit, as viewed by SEM (×66). Note the gradual transition from the small, densely packed cells of the flavedo to the loosely attached, highly degenerate albedo. Also note the structure of the oil gland

Renewed growth of the albedo and thickening of the peel occur in certain cultivars during fruit maturation and senescence.

The pulp, which is the edible portion of the fruit, consists of segments, the ovarian locules, enclosed in a locular membrane and filled with juice sacs (sometimes called juice vesicles). Development of the juice sacs has been followed in detail by Schneider (1968). Juice sacs are initiated at about bloom, appearing at first as dome shaped protrusions from the locular membrane into the locules (Figure 4.9). Development of the domes into juice sacs occurs through apical meristem activity and subsequently by an obese mass of meristematic tissue giving rise to the body of the sac. In the mature fruit, juice sacs appear as elongated, mostly spindle-shaped multicellular structures, projecting from a stalk implanted in the periphery of the segment toward the central axis, where the

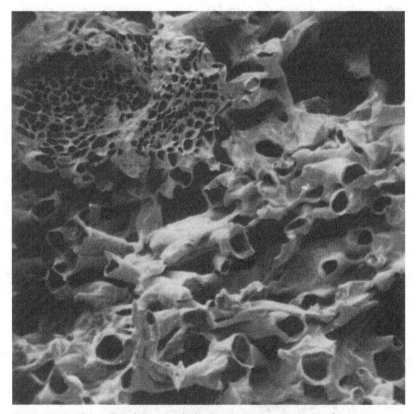

Figure 4.12 Structure of the albedo (= the white, spongy layer of the peel) from mature Valencia orange (*C. sinensis*) fruit, as viewed by SEM (×270). The 'holes' are cross sections of albedo cell lobes. A cross section of a vascular bundle appears at the upper left

seeds are found (Figure 4.13). The overall structure of the mature fruit
with its vascular systems is shown schematically in Figure 4.14.

Juice sacs are the ultimate 'sink' organ of citrus and their development
presents intriguing physiological problems. The fact that the juice sacs are
not connected to the vascular system which provides water and assimilates to the fruit has long been recognized by students of fruit morphology
(Schneider, 1968). Recent studies by Koch and coworkers (Koch *et al.*
1986, Koch & Avigne, 1990) have confirmed and extended these observations. Neither juice sacs nor their stalks (Figure 4.13) nor the segment
epidermis to which the stalks are attached show any differentiation of
transfer organs. Thus, juice sacs must obtain their supply over long
distances (up to 3 cm) of postphloem, through nonvascular cell-to-cell
transport. Labeling experiments indicate that the entry of water and
assimilates into juice sacs is extremely slow. It is still not clear whether the
cell-to-cell transfer of assimilates takes place via symplastic (i.e. plasmo-
desmata) or apoplastic (i.e. cell wall) routes. The biophysical mechanisms

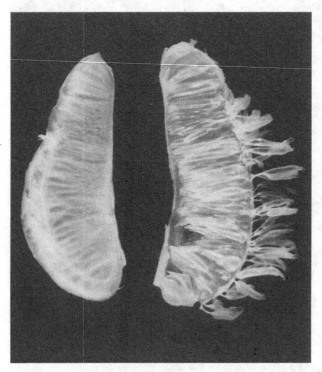

Figure 4.13 Segments of mature pummelo (*C. maxima*) fruit. The
segment to the left is intact. The skin of the segment to the right has
been removed to show the juice sacs (*c.* ×0.8)

of water and solute uptake by the growing fruit as well as the biochemical reactions involved need further elucidation.

Fruit development

In a classical study, Bain (1958) divided the development of Valencia orange fruit into three major stages: cell division (I), cell enlargement (II) and fruit maturation (III). Bain's division seems appropriate for most citrus fruits, although times and duration of developmental stages may vary, according to cultivar, climate, etc. (Figure 4.15). In Figure 4.15 the hatched areas marked between stages I/II/III are meant to indicate that it is impossible to draw an exact borderline between these somewhat-overlapping stages.

The cell-division period (stage I) may be assumed to commence at fruit set, immediately following anthesis. However, differences in fruit size between leafy and leafless inflorescences are already evident in the ovaries prior to anthesis (Guardiola & Lazaro, 1987; Hofman, 1988). The size of the ovary also varies as a function of the number of flowers per tree

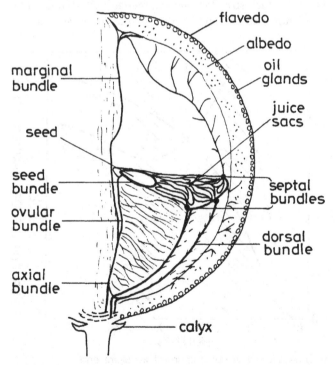

Figure 4.14 Schematic drawing of a mature citrus fruit emphasizing the vascular arrangement

(Guardiola *et al.*, 1984). These differences in size probably reflect differences in cell division during floral development, indicating that fruit development actually begins before anthesis, a view expressed long ago by Nitsch (1953).

Cell division appears to terminate in all fruit tissues, except the outermost flavedo layers and the tips of juice sacs, within five to ten weeks after bloom. The increase in fruit size during stage I is mainly due to growth of the peel, consisting of cell division, but there is already a component of cell enlargement. The peel reaches its maximum width at or soon after the end of stage I (Figure 4.15); this has been shown repeatedly for oranges (Bain, 1958; Goren & Monselise, 1964; Holtzhausen, 1982), grapefruit (Herzog & Monselise, 1968) and mandarins (Kuraoka and Kikuki, 1961). Peel volume increases nevertheless somewhat further during stage II.

Stage II, the cell-enlargement phase, may also be envisaged as the pulp growth stage. Juice sacs enlarge and fill the locules (segments) quite early, with their juice and sugar content increasing mainly towards the end of this stage. The rapidly expanding pulp exerts considerable pressure outwards on the peel, which stretches, getting increasingly thinner. The shape of the oil glands also changes as the season progresses, due to

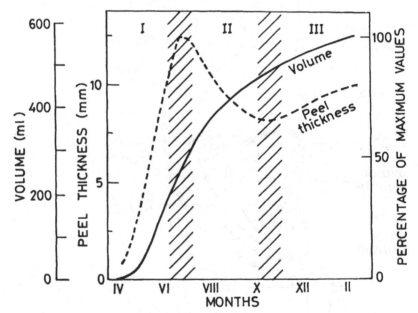

Figure 4.15 Fruit growth and development: growth in volume and peel thickness. I, II, III indicate developmental stages according to Bain (1958). Modified from Monselise (1986)

stretching of the peel (Holtzhausen, 1982). Fruit splitting, which is quite widespread among thin-peeled mandarin and orange cultivars (Figure 4.16), is believed to result from excessive pressure of the developing pulp on the thin, over-stretched peel (Goldschmidt *et al.*, 1994).

Stage III is known as the fruit maturation and ripening phase (Bain, 1958). In fact, fruit growth continues during this stage as well, the growth rate depending to a large extent on climatic conditions (Figure 4.17). As mentioned, renewed growth and thickening of the peel may be observed during stage III (Figure 4.15), particularly under warm, growth-favoring conditions (Kuraoka, 1962; Herzog & Monselise, 1968; Reuther, 1973). Pulp growth, on the other hand, almost stops at this stage in certain cultivars (e.g. Satsuma), leading to formation of cracks between the peel and the pulp. When combined with lysis and degradation of the albedo these cracks develop into large, hollow air spaces, a condition known as 'puffing' (Kuraoka, 1962).

Studies of citrus fruit development have generally focused on a specific cultivar. However, when the whole spectrum of citrus fruits is considered attention has to be paid to the diversity of peel/pulp ratios in mature fruit (Figure 4.18). On one extreme are certain types of citron (*Citrus medica*), which lack a fleshy pulp altogether. The segments are thick-walled, empty locules containing only the seed. All the rest of the fruit cross-sectional

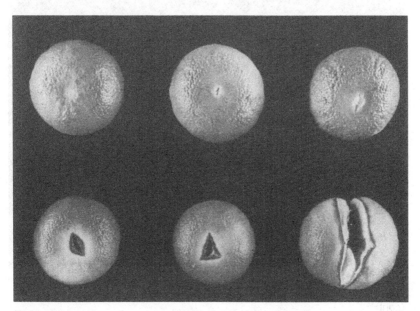

Figure 4.16 Development of the fruit splitting phenomenon in Murcott tangor

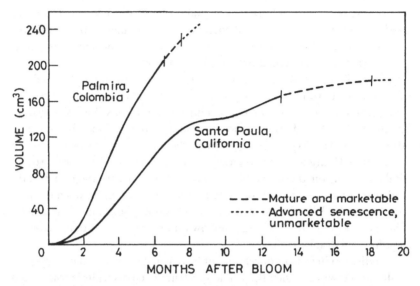

Figure 4.17 Schematic presentation of typical Valencia orange growth curves in two widely different climatic situations. In addition, periods of immaturity, market maturity and advanced senescence are approximated (modified from Reuther, 1973)

Figure 4.18 Cross sections of mature fruit from thick-peeled, pulpless Yemenite citron (*C. medica*) (left) and thin peeled Murcott tangor (right)

area is occupied by the albedo and by the massive central axis. At the other end are thin-peeled mandarins. Here, the pulp predominates while the albedo had degenerated and almost disappeared. Between these extremes are all other kinds of citrus – pummeloes, grapefruit and oranges represent, in a decreasing order, different peel/pulp ratios. This is undoubtedly a genetically controlled trait which might, to some extent, be mediated by plant hormones.

Fruit abscission

The term 'abscission' has been assigned to the shedding of plant organs, commonly known as fruit (or leaf) drop. Abscission became only recently acknowledged as a genetically programmed, hormonally controlled developmental process (Osborne, 1989). In citrus, which is an evergreen, most research efforts have been focused on fruit abscission. Leaf abscission also received considerable attention, however, and citrus leaf explants were among the first detached organ systems used in physiological studies of abscission (Addicott et al., 1949).

Two major kinds of fruit abscission may be discerned during fruit development and each seems to have a specific role. Fruitlet abscission (which in many cultivars commences, in fact, during bloom) is a self-thinning mechanism which adjusts the number of fruit to the tree's bearing potential. The shedding of mature fruit, on the other hand, may be regarded as part of the evolutionary seed-dispersal program. Both abscission phases have practical implication. Fruitlet thinning is practiced in cases of excessive fruit set (e.g. Murcott tangor and other alternate bearing cultivars), or where the reduction of fruit numbers will pay off through increase in fruit size (Goldschmidt and Monselise, 1978). The abscission of mature fruit, often called 'preharvest drop', occurs in various cultivars, especially during cool, wet winter months. It is assumed to result from decline in the stream of native auxin reaching the fruit. Indeed, relatively low concentrations of synthetic auxins have long been in use for control of this type of abscission (see Chapter 5, Table 5.9).

Additional waves of fruit drop occur in certain cultivars between the end of fruitlet drop and fruit ripening. The phenomenon was described in navel orange and is apparently related to the presence of the secondary fruit (the navel) (Lima et al., 1980). Fruit splitting, which occurs in various C. reticulata cultivars and hybrids during late summer–early autumn, also results in massive fruit drop. These abscission waves appear to be associated with certain mechanical injuries which, presumably, result in the production of ethylene.

Considerable efforts have been invested in the search for chemicals which lead to 'fruit loosening', to facilitate mechanical harvesting of fruit (Wilson *et al.*, 1982). Most of these chemicals directly or indirectly cause ethylene evolution, which unfortunately results also in severe leaf drop. This is a major drawback which prevents the adoption of these chemicals for orchard use.

Abscission takes place in predetermined 'abscission zones'. Abscission of young fruit occurs between the fruit peduncle and the subtending branch. Later on, an increasing number of fruit abscise at the calyx, as do mature fruit. The abscission process involves anatomical and biochemical changes which culminate in the physical separation between the abscising organ and the parent plant. The sequence of events, as summarized for citrus by Goren (1993), begins with cell division and elongation, followed by disappearance of starch grains, collapse of the cell walls, which turn into a gelatinous mass, and dissolution of the middle lamella. The hydrolytic enzymes, cellulase and polygalacturonase, are responsible for the breakdown of the cell wall components. Ethylene enhances the abscission process, whereas auxin slows it down. The interplay of ethylene and auxin in the regulation of abscission is complex, including effects of ethylene on auxin transport and conjugation. The hormonal control of abscission probably involves gene activation but the details await elucidation.

Effect of climate on fruit development and quality

Citrus is an evergreen, subtropical crop and low temperatures are the main factor restricting its geographical distribution. Frost and freezing damage the fruit and, when lasting long enough, may kill the trees. Even at the milder, non-damaging range, temperatures present major limitations for vegetative growth as well as for fruit development and maturation. Little growth occurs in all citrus tree organs below 13 °C. However, in areas where temperatures rise above that minimum only for a relatively short summer period, both vegetative and reproductive development may also be very much restricted.

Table 4.1 summarizes temperature, heat unit accumulation and rainfall data for several citrus growing areas around the world. Annual averages tell, of course, only part of the story; monthly or even daily fluctuations reveal subtle changes which are also physiologically meaningful. The cumulative heat-unit method which is commonly used for prediction of harvest dates is also rather crude, since the response of fruit at different developmental stages to temperature varies considerably (Newman *et al.*, 1967). Still, heat-unit data seem to correlate reasonably

Table 4.1 *Annual mean maximum, mean minimum, range and average temperatures, annual heat unit accumulation[1], annual rainfall and brief definition of climate in various citrus-growing regions*

Location	Latitude	Temperature (°C)				Heat units	Rainfall (mm)	Definition of climate
		Maximum	Minimum	Range	Average			
Rehovot, Israel	32° N	26.2	14.0	12.2	20.1	2595	580	Subtropical
Valencia, Spain	39½° N	20.8	12.3	8.6	16.5	1626	397	Mediterranean, cool
Wakayama, Japan	34° N	21.3	11.8	9.5	16.6	1951	1808	Maritime, cool
Kerikeri, New Zealand	37° S	20.2	9.7	10.5	15.0	896	1656	Maritime, cool
Nelspruit, South Africa	25½° S	26.7	13.7	13.0	20.2	2607	812	Semitropical
Orlando (Florida), USA	28½° N	28.2	16.7	11.5	22.4	3465	1339	Semitropical
Santa Paula (California), USA	34½° N	24.2	7.6	16.6	16.2	1258	317	Subtropical, cool–dry
Palmira, Colombia	3½° N	29.9	18.0	11.9	23.9	3918	1010	Tropical

[1] Calculated as the annual sum of the (average monthly temperature − 13) × (no. of days per month).

Compiled from data from Reuther (1973).

well with rates of fruit development and maturation (Reuther, 1973). Differences in date of maturation between 'early' and 'late' cultivars are believed to reflect differences in heat unit requirements – late cultivars require a larger sum of heat units. Thus, areas with low annual heat sums are inclined to grow early ripening cultivars (Clementines, Satsumas).

While Table 4.1 shows the climatic data, Figure 4.17 demonstrates the ensuing differences in fruit development between a subtropical, cool dry location (Santa Paula, California) and a tropical location (Palmira, Colombia). Under the high temperatures prevailing in the tropics fruit development is fast and fruits get very large. In California fruit development is much slower. Fruit growth stops during cool winter months and resumes again at a low rate during spring, but the final size of fruit is considerably smaller than that obtained in the tropics. The heat unit requirement for maturation of Valencia orange is fulfilled in tropical Colombia within 6.5 months, while in cool California about double this time is required. Correspondingly, tropical Valencia fruit remain 'mature and marketable' only for a short time, followed by rapid senescence. The California fruit, on the other hand, has an extended period of maturity during which the fruit may be harvested and marketed. Less extreme differences in heat-unit accumulation, such as found between mild, coastal and inner, desert-like locations (e.g. in Israel or California) are nonetheless sufficient to produce several weeks' delay in maturation dates (Herzog and Monselise, 1968; Reuther, 1973).

Climate affects fruit quality as well. Rind color is a major problem in the tropics – warm temperatures interfere with the loss of chlorophyll as well as with the build up of carotenoids. Thus, fruit in the tropics stay greenish and pale; oranges and mandarins, in particular do not attain their attractive rind color. Cool temperatures, on the other hand, enhance the desired color changes. The autumn decline in air and soil temperatures marks the onset of color changes in subtropical regions (Young and Erickson, 1961). This view is supported by a large number of field observations and could also be simulated in controlled greenhouse experiments (Reuther, 1973).

Combinations of high temperature and high humidity result in tender, rapidly senescing fruit which has low storage potential and is highly susceptible to peel blemishes. A comparison between coastal and desert grown fruit in California has shown that the peel of fruit developing under the drier climate has a lower water content and is not so tender, presumably due to the hardening effect of moisture stress (Monselise and Turrell, 1959).

Internal quality is also affected by climate. Fruit developing in a hot, tropical climate tends to have a high total soluble solids content, which is an advantage for the processing industry. On the other hand, these fruits are low, often very low, in acid, resulting in poor edible quality. Thus, the somewhat cooler, subtropical areas are preferable for production of oranges and mandarins for the fresh fruit market.

Maturation, ripening and senescence of citrus fruit

Ripening of citrus fruit is quite different from that of most other fruits. Ripening of fruits like apple, avocado, tomato and banana involves rather abrupt changes in fruit texture and composition. In citrus fruit such changes are rather limited and take place in a slow and gradual manner. Citrus fruit approaching maturation does not contain starch and must, therefore, achieve internal maturity on the tree, prior to harvest. The biochemical changes occurring in avocado, tomato, etc. appear to be intimately related to the climacteric rise in respiration and ethylene evolution (Theologis *et al.*, 1992). Citrus fruits, on the other hand, are 'non-climacteric' – respiration declines slowly throughout the later stages of fruit development and ethylene evolution of the mature fruit is extremely low (Aharoni, 1968; Eaks, 1970; Goldschmidt *et al.*, 1993). Use of the term 'ripening' with regard to citrus is often refuted on these grounds and the term 'maturation' is preferred.

Structural and physiological differences between peel and pulp of citrus fruit have already been pointed out in the foregoing discussion of fruit development. During maturation peel and pulp behave in most respects as separate organs, although some coordination does exist. Figure 4.19 describes some of the major changes occurring in peel and pulp of Shamouti orange during maturation. The changes which take place during maturation of the flavedo are comparable to the senescence of other chlorophyllous tissues, as revealed mainly in transformation of the chloroplasts into chromoplasts (Goldschmidt, 1988). The decline in rind chlorophyll takes several months and the onset of carotenoid accumulation almost coincides with the disappearance of chlorophyll. Just prior to their build-up the carotenoids go through a 'trough' which marks the transition from carotenoids of the photosynthetic chloroplast to the intensely colored carotenoids of the chromoplast (Eilati *et al.*, 1969; Gross, 1987).

Maturation of the pulp is characterized by gradual changes in juice content and in some of its constituents. On one hand, there is a decline in titratable acidity (TA) brought about by decomposition of citric acid, which is the principal organic acid of citrus juice (Monselise, 1986). On the other hand, there is an increase in sugars, usually expressed as total soluble solids (TSS) (Figure 4.19). With acidity declining and sugars increasing towards maturation, the TSS/TA ratio is extremely sensitive and is commonly used, therefore, as a 'maturity index'. The arrow in Figure 4.19 indicates a TSS/TA ratio of 8 which has been reached, in this case, at about the same time that chlorophyll disappeared. This is not the rule, and peel coloration of different cultivars may often precede or lag behind internal maturity. Degreening with ethylene is practiced in the latter case, mainly with early cultivars (Grierson *et al.*, 1986).

The increase in the percentage of expressible juice (juice content, Figure 4.19) is at least partly due to release of water within the pulp tissue,

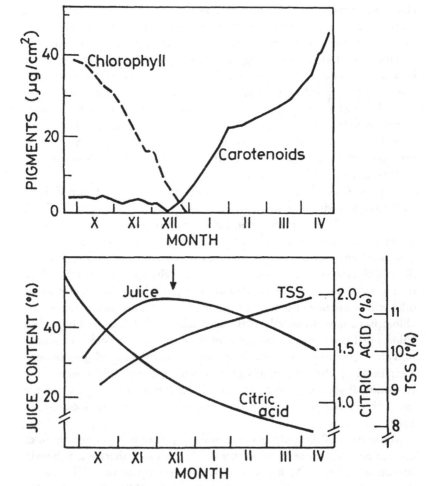

Figure 4.19 Major biochemical changes occurring during maturation of Shamouti orange (*C. sinensis*) (see text)

also occurring postharvest. Maturity standards are legally enforced in several different countries, varying for different species and cultivars. For oranges, grapefruits and mandarins they usually include an accepted range of TSS/TA ratios, a minimum juice content and an acceptable peel color. Juice content is the only maturity parameter used for lemons and limes, which are commonly used as a non-sweetened acid ingredient (Soule and Grierson, 1986).

Fruit softening, which is a dominant feature of ripening in most climacteric fruits, does not play a significant role in the maturation of citrus fruits. Although both peel and pulp of citrus fruit are rich in polyuronides (pectins), their decomposition into smaller, soluble subunits is very slow, with the exception of certain mandarins, which undergo a more pronounced softening. The postharvest reduction in fruit firmness is largely due to loss of water, mainly from the pulp, leading to shrivelling and deterioration of fruit appearance and quality. Real 'softening' is brought about by several pathogenic postharvest diseases involving massive secretion of cell-wall degrading enzymes (Eckert and Eaks, 1989).

Once the fruit has reached maturity it can either be harvested or 'stored' on the tree. Fruit can be held on the tree for rather long periods of time, provided that insects (such as the Mediterranean fruit fly, *Ceratitis capitata*) are under control. Preharvest fruit drop may be a problem under cool–wet winter conditions, particularly in mandarin cultivars; spraying with low concentrations of synthetic auxins is recommended in these cases (Coggins and Hield, 1968; Goren, 1993). Changes in internal quality nonetheless set a limit to the delay of harvest. Cultivars vary greatly in this respect – whereas grapefruits and Valencia oranges retain high quality for months, Shamouti oranges, and even more so soft mandarin-like cultivars, deteriorate much faster. Loss of quality is sometimes associated with physiological disorders such as 'granulation' (Grierson, 1986). On the other hand, delay of harvest may enable a further increase in TSS content, which might be advantageous for the processing industry (Halpern and Zur, 1988).

Citrus fruits have a considerable postharvest storage potential but varietal differences exist in this case as well. Grapefruits and Valencia oranges can be stored for three to five months and green lemons even longer, but many easy peeling, mandarin-like cultivars cannot be stored for longer than a few weeks. Storage temperatures also differ – grapefruits are sensitive to chilling injury and should be stored at 10 to 16 °C; oranges and mandarins are stored at lower temperatures. Controlled atmosphere is not used with citrus fruit. Seal packaging of individual fruit with polyethylene film has been tried but is not yet widely adopted (Grierson and Ben-Yehoshua, 1986).

Citrus fruit are the most widely exported fresh fruit. Shipment by sea and prolonged marketing periods have necessitated the development of sophisticated postharvest treatments, to ensure extended fruit viability and protection from pathological deterioration. Packing houses have become, therefore, an important station on the way from the orchard to the fresh fruit market. In the packing house the fruit is first thoroughly rinsed and scrubbed and subsequently waxed, to replace the natural wax which had been removed during the cleaning process. In the packing house the fruit is also supplied (usually through the wax) with fungistatic chemicals for the control of postharvest decay diseases and with growth regulators: 2,4-dichlorophenoxyacetic acid (2,4-D) to prevent abscission of the pedicel at the calyx, and gibberellin A_3 (GA_3) to increase viability (Coggins and Hield, 1968; Monselise, 1979; Davies, 1986; see also Chapter 5). GA_3 also increases fruit resistance to certain pathological diseases (Coggins et al., 1994). Growing concern in recent years over the amount of toxic residues in fruit has prompted efforts to develop physical and biological treatments which should replace, or at least reduce, the amounts of agrochemicals applied to fruit. This trend will probably become more dominant in years to come.

The control of fruit development and maturation

Various systems interact in the regulation of fruit growth and numerous factors turn out to be limiting during the course of fruit development. An ample supply of water is a prerequisite for all stages of fruit development. Water stress can be particularly dangerous during fruit set, leading to a massive drop of fruitlets (Monselise, 1986). Increase in size and juice content are also largely dependent upon the availability of water (Marsh, 1973). Mineral nutrition, with its complex elemental interactions, is important as well. Potassium seems to play a special role in fruit development – potassium deficiency reduces fruit size (Chapman, 1968; Embleton et al., 1973b; Du Plessis and Koen, 1988) and potassium sprays are used to strengthen the peel (Embleton et al., 1973b) and increase fruit size. The supply of carbohydrates has also been suggested as a limiting factor for fruit set and enlargement. Girdling at full bloom increases fruit set (Monselise et al., 1972; Erner, 1988), presumably by provision of more photosynthate to the young fruitlets (Schaffer et al., 1985). Girdling during the fruit enlargement stage increases fruit size (Cohen, 1984; Fishler et al., 1983), probably by eliminating competition for photosynthate by the root system and other growing organs.

Major significance has been assigned to plant growth substances in the control of fruit set, growth and maturation since the classical work of

Nitsch (1953). Attempts to correlate the changes in levels of endogenous growth substances with growth rates of fruit components have not always been successful but have, nonetheless, provided interesting clues. Complementary evidence has been derived from the responses of fruit to exogenous applications of growth regulators.

It has been clear from the outset that various groups of growth substances are involved in the control of fruit development. There is a difficulty, however, in assigning specific roles to each substance. It has been shown time and again that young, vigorously growing plant organs contain high levels of the growth-promoting substances, whereas levels of abscisic acid (ABA) and other growth inhibitors increase towards maturation and senescence. Citrus fruits are no exception – young fruitlets have relatively high levels of auxins, gibberellins and cytokinins while the flavedo of mature fruit contains large amounts of ABA (cf. Goldschmidt, 1976). The advent of modern techniques for identification and quantification of endogenous growth substances and their derivatives has revealed the extreme complexity in this domain, particularly of the complement of endogenous gibberellins (cf. El-Otmani et al., 1995).

The concept of the developing seeds as the principal source of growth substances for the developing fruit (Nitsch, 1953) does not seem to be readily adaptable to citrus fruits, many of which are absolutely seedless or nearly so (see Chapter 6). The peel has been suggested, instead, as a center of hormonal regulation in the developing citrus fruit (Monselise, 1978). Nevertheless, an exogenous supply of gibberellins has repeatedly been shown to increase fruit set in Clementine and other self-incompatible C. reticulata hybrids (Krezdorn, 1982; El-Otmani et al., 1994), suggesting a role for the fertilized ovule in the provision of endogenous gibberellins for fruit set.

Synthetic auxins (such as 2,4-dichlorophenoxyacetic acid and related compounds) bring about an increase in fruit size when applied to young fruitlets (Coggins and Hield, 1968; Monselise, 1979; Guardiola and Lazaro, 1987), indicating a role for auxins in fruit growth, which at this early stage involves primarily peel growth. The responsiveness to auxins diminishes rapidly, however, and treatments given after the fruit has reached about a third of its final diameter are usually without effect (see also Chapter 5).

Rough thick-peeled fruit, which appear occasionally on juvenile Shamouti orange trees, have high levels of endogenous gibberellin-like substances (Erner et al., 1975). Application of growth retardants, which arrest the biosynthesis of gibberellins, reduces the incidence and extent of peel roughness (Erner et al., 1976). These results point to the role of cytokinins and gibberellins in peel development. Further support for the role of gibberellins in peel growth derives from the fact that exogenous gibberellins reduce the incidence of creasing and of related peel-deterioration phenomena (Embleton *et al.*, 1973a; Monselise *et al.*, 1976; Coggins and Henning, 1988).

Growth substances also play an important role in the control of fruit maturation. This, again, concerns mainly the peel – internal quality and juice content are seldom affected by growth regulators (Coggins and Hield, 1968; El-Otmani and Coggins, 1991). Chloro–chromoplast trans-formation, as revealed in peel pigment changes, is the major physiological trait affected by growth substances. While the role of endogenous ethylene in the triggering of citrus fruit maturation requires further, unequivocal proof (Goldschmidt *et al.*, 1993a), exogenous ethylene markedly acceler-ates the senescent pigment changes of the peel, particularly the loss of chlorophyll. Gibberellins and cytokinins, on the other hand, delay the senescent color changes (cf. Goldschmidt, 1988). Chloro–chromoplast transformation is, to some extent, reversible – the phenomenon is known as 'regreening'. Regreening consists of renewed build up of chlorophyll and chloroplast membranes and the photosynthetic activities are also partly restored (Thomson *et al.*, 1967; Saks *et al.*, 1988). Mature Valencia orange fruit held on the tree undergo regreening during the summer and the process is enhanced by gibberellins (Coggins and Lewis, 1962).

Environmental, nutritional and hormonal signals appear to be involved in the control of chloro–chromoplast interconversions, as summarized in Figure 4.20 (Goldschmidt, 1988). Although the experimental evidence is incomplete, the following hypothesis seems to account for most observa-tions. As long as temperatures permit root growth, root hormones (gibberellins, cytokinins) and other nitrogenous substances reach the canopy and delay the senescent color changes. When autumn sets in, temperatures drop, root growth stops, the level of root substances declines and peel senescence ensues. The renewal of root growth during spring re-elevates the level of root hormones, thereby leading to regreening. In the tropics, high temperatures prevail all year round, root growth occurs uninterruptedly and root substance levels are always sufficiently high to interfere with fruit coloration. Plant growth substances are thus inti-mately involved in all stages of fruit set, development, maturation and senescence.

Fruit composition

Table 4.2 lists several classes of native constituents of citrus fruits, and also gives some estimates of their concentrations in peel and juice of orange and lemon. The significance of some of these compounds is related to citrus fruits' edible and nutritional qualities, while others are important by-products of the processing industry.

The soluble sugar pool contains mainly glucose, fructose and sucrose. In orange, grapefruit and mandarin juice the amount of sucrose exceeds that of fructose and glucose. Pulp sucrose levels increase markedly towards ripening, reaching in certain mandarins 15 to 18% of fresh weight (Tzur, 1994). In the peel, the soluble sugar pool often contains more reducing sugars (glucose, fructose) than sucrose. In juices of lemon and lime the amounts of sucrose are minimal.

Ascorbic acid, better known nutritionally as vitamin C, is biochemically related to sugars although its precise biosynthetic route in citrus has not been elucidated. Citrus fruits have been known for a long time as a major dietary source of ascorbic acid. As early as the seventeenth century citrus fruits have been found to prevent the scurvy (scorbutus) disease. A teaspoonful of citrus juice was often included, therefore, in the daily ration of seamen (Sinclair, 1984). Citrus fruit peel has a higher ascorbic acid content than the juice. Fruit exposed to sunlight have a significantly higher ascorbic acid content, as well as a higher total soluble

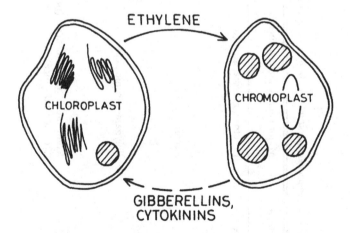

Figure 4.20 Schematic diagram of chloroplast–chromoplast transformation and the major regulatory systems involved

Table 4.2 *Classes of compounds present in citrus fruit peel and juice*

Class of compound	Major chemical constituents	Range of concentrations (mg g^{-1} fresh wt. or ml juice)	Key references
Soluble sugars	Fructose, glucose, sucrose	67.4–83.4 (orange peel) 87.8–110.6 (orange juice) 32.1–58.6 (lemon peel) 11.2–25.7 (lemon juice)	Ting & Attaway (1971); Sinclair (1984)
Vitamin C	Ascorbic acid	1.30–2.20 (orange peel) 0.40–0.60 (orange juice) 1.60–2.30 (lemon peel) 0.50–0.80 (lemon juice)	Eaks (1964); Sinclair (1984)
Pectin	Polygalacturonic acid	40–82 (orange peel) 0.5–0.6 (orange juice) 27–62 (lemon peel) 0.3–0.8 (lemon juice)	Sinclair (1984)
Organic acids	Malic acid, malonic acid (peel), citric acid (juice)	3.1–4.9 (orange peel) 14–22 (orange juice) 2.4–3.6 (lemon peel) 58–62 (lemon juice)	Sinclair (1984); Sasson & Monselise (1977)

Flavonoids, flavanone, glucosides	Hesperidin (orange, mandarin, lemon) Naringin (grapefruit, pummelo)	5.0–15.0 (orange peel) 0.20–0.22 (orange juice) 1.5–2.0 (lemon whole fruit) 0.2–0.4 (lemon juice)	Goren (1965); Sinclair (1984)
Carotenoids	Violaxanthine, luteo-xanthines (orange) Phytofluene, β-carotene, cryptoxanthine (lemon)	0.050–0.120 (orange peel) 0.006–0.015 (orange pulp) 0.0014–0.0021 (lemon peel) 0.0006–0.0011 (lemon pulp)	Gross (1987)
Anthocyanins	Cyanidin-3 glucoside	(blood orange)	Gross (1987)
Essential oils	d-limonene	4.3–9.0 (orange peel) 5.9–7.7 (lemon peel)	Sinclair (1984)
Limonoid teriterpene derivatives	Limonin	0.0086–0.0192 (orange juice)	Sinclair (1984); Ting & Rouseff (1986)

solids content. The ascorbic acid content does not change much after harvest, even during several months' storage. Processing of fruit may involve considerable losses, unless special precautions are exercised (Sinclair, 1984).

Pectins are major cell-wall components of fleshy fruits. Pectins are large, complex carbohydrate macromolecules, composed of partly methylated polygalaturonic acid backbones and considerable amounts of other sugar residues. Various methods for extraction and separation of pectins from other cell-wall components have been developed. Extensive research has been conducted on the pectic materials of citrus fruits (Sinclair, 1984). Citrus pectins have a high galacturonic acid content. Citrus fruit, particularly the peel, serve as an important raw material for production of high-quality commercial pectin. Pectin is used in the food industry, mainly as a jellying agent.

Acidity is a major determinant of fruit taste and edibility. Organic acids play a central role in fruit metabolism (Ulrich, 1970). Malic acid in apple, tartaric acid in grape and citric acid in *Citrus*, all show a distinct peak at the midst of the growth period, followed by a descent towards ripening (Monselise, 1986). The citric acid of citrus juice is probably produced by a side cycle, coexisting with the tricarboxylic acid cycle. Accumulation of citric acid has been proposed to be brought about by the high concentration of citramalate, which blocks aconitase activity (Wallace *et al.*, 1977). Citrus fruits vary greatly in their acid content; even cultivars of the same species (e.g. orange, lime) show extreme differences. The genetics of the acidless trait has been studied in several *Citrus* species and found to be complex (Cameron & Soost, 1979). The biochemical background of these differences in acidity is as yet poorly understood (Wallace *et al.*, 1977).

While citric acid is the main acid component of juice, with malic acid coming next, malic and malonic are the major acids in the flavedo and albedo (Sasson & Monselise, 1977). The total acid content of the peel is much lower, however, than that of the juice (*c.* 5%). Malonic acid appears to increase during rind senescence and off the tree (Monselise, 1986).

The flavonoids are an important class of plant secondary metabolites, belonging to the broader family of plant phenolics. The flavonone glucosides of citrus have been subject to numerous studies over the years (Kesterson & Hendrickson, 1953; Goren, 1965; Sinclair, 1984). Considerable progress has been achieved recently in their analysis (Castillo *et al.*, 1992) and their biosynthetic pathway, including some of the key enzymes involved, has been elucidated (Hasegawa & Maier, 1982; Bar Peled *et al.*, 1993). Hesperidin and naringin are known to be the major flavanone glucosides of orange and grapefruit, respectively. The overall distribution of flavonoids in *Citrus* species is more complex, however, and attempts have been made to use it as a chemotaxonomic tool (Tatum *et al.*, 1978). Citrus flavonoids accumulate in young, rapidly developing organs – attaining up to 75% of the dry weight of young fruits of approximately 1 cm diameter (Kesterson & Hendrickson, 1953). No specific role can be assigned so far to the flavonoids in the physiology of citrus. Citrus flavonoids have aroused much interest due to their organoleptic properties. Whereas hesperidin is tasteless, naringin is extremely bitter and as such responsible for the bitterness of grapefruit. Neohesperidine dihydrochalcone is an intense artificial sweetener (Horowitz & Gentili, 1986) which can be produced from naringin or, even more easily, from neohesperidine, which is a natural flavanone glucoside of *Citrus aurantium* (Castillo *et al.*, 1992).

The carotenoids of citrus fruits have been extensively investigated, particularly by Gross and coworkers (Gross, 1987 and references therein). The *Citrus* genus is remarkable for producing the largest number of carotenoids found in fruit. About 115 different pigments have been reported, including a large number of isomers, some of which might be formed during isolation.

Each species and hybrid has a characteristic carotenoid complex which is responsible for its typical color; peel (flavedo) and pulp pigments reveal certain differences. In yellow citrus fruits (pummelo, grapefruit, lemon, lime) the total amount of carotenoids is low and most of these belong to the colorless carotenoids. Thus, in white Marsh seedless grapefruit the colorless phytoene and phytofluene account for 74% of the carotenoids of the flavedo. The accumulation of these colorless precursors is the result of a genetically determined metabolic block which hinders further dehydrogenation steps leading to the colored carotenoids. In pink and red grapefruit cultivars lycopene and β-carotene are the major carotenoids, particularly in the pulp, indicating that the genetic block had been removed. Lycopene is also responsible for the rapid formation of red color in citrus fruit treated with tertiary amine bioregulators (Yokoyama & Keithley, 1991).

Orange-colored citrus fruits (orange, sour orange, mandarin) contain relatively large amounts of a complex mixture of carotenoids. Some of these (e.g. cryptoxanthine, β-citraurin) appear in small amounts but have a high tinctorial value. The molecular regulation of this extreme biochemical diversity within the *Citrus* genus is still greatly unknown.

Blood oranges, on the other hand, owe their color to a different class of pigments – the anthocyanins. Like most anthocyanin-containing fruits, blood oranges also develop the desired, intense coloration in cooler regions. This is presumably the reason for the success of blood oranges under the cool maritime climate of Sicily.

Essential oils are volatile, fatty plant constituents contained in the oil glands found in most citrus organs, particularly in the flavedo. The essential oils are chemically composed of terpenes, aldehydes, alcohols, acids and hydrocarbons and rarely exist as esters of the ordinary fatty acids. Lemon oil was found by Kesterson *et al.* (1971) to contain more than 100 chemical constituents, 68 of which have been definitely identified. The monoterpene, *d*-limonene, is the principal constituent (60–90%) of fruit essential oils, whereas leaves have other monoterpenes as their major constituents. Analyses of leaf essential oils have been used for chemotaxonomy of *Citrus* species. Citrus essential oils are widely used for cosmetic and pharmaceutical purposes, as well as flavoring agents in the food and beverage industry. Records of commercial lemon oil production in Italy date back more than four centuries. The Italian lemon oil, produced by the old, hand-pressing method has been renowned for its high quality. In Sicily and other lemon-growing areas of southern Italy, a considerable portion of the crop, at times even most of it, was diverted to the essential oil industry. There are two types of commercial essential oils: coldpressed oil and distilled oil. The production method affects the yield and chemical composition of the oil. Commercial yields of lemon oil range between 0.54 and 0.78% of fruit weight (Sinclair, 1984).

About 30 limonoid triterpene derivatives have been identified in *Citrus* species. Limonin is the major member of this group. Limonin is the principle bitter compound of citrus juices and as such has been subject to extensive research (Maier *et al.*, 1980; Hasegawa & Maier, 1982; Ting & Rouseff, 1986). In intact fruit, limonin is present in the monolactone form, which is the nonbitter precursor of limonin (Maier & Beverly, 1968). This explains the delayed development of bitterness in citrus juices. Considerable efforts have been put into the development of enzymatic debittering techniques. Similar approaches have been undertaken also with regard to the naringin bitterness of grapefruit juice (Hasegawa & Maier, 1982).

The compounds listed in Table 4.2 represent some of the best known constituents of citrus fruits. Citrus fruits contain, of course, many additional native compounds which may turn out to be of interest for human use in the future.

Citrus productivity

Productivity is the final outcome of a long chain of developmental events. 'Failure' in any one of these steps will normally result in lack of fruit. Having gone through the various stages of flowering and fruiting in previous sections it may be worthwhile to take a look at citrus productivity at the overall, whole-tree level.

Figure 4.21 represents in a schematic way the dynamic processes involved in citrus crop production. Major processes are spelled out in rectangles along the time axis, while the two curves show the diminishing number of reproductive units with the increasing weight of the individual fruit unit. Keeping this general scheme in mind we shall now discuss the critical stages of the fruiting process.

Number of flowers and type of inflorescence

Flower formation is certainly a critical step, since an absolute lack of flowers would preclude yield formation altogether. In reality, however, flower number is rarely a limiting factor (0.5 to 3 $\times$ 10⁵ flowers per mature tree have been recorded), as the large majority of flowers drops anyway and only very few persist on the tree to become mature fruit. Only in a few cases (such as young, partially juvenile trees, or alternate-bearing trees

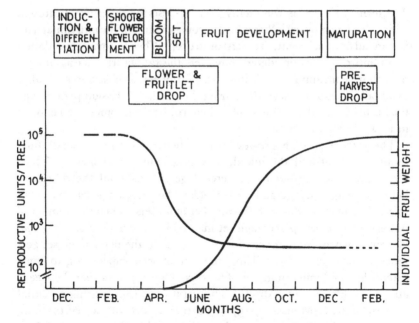

Figure 4.21 Schematic description of the annual cycle of crop production in a Citrus tree. Rectangles represent major processes along the time axis. Curves show the diminishing number of reproductive units (on a logarithmic scale) as against the increase in weight of the individual fruit unit. Modified from Goldschmidt and Monselise (1978)

during their 'off' year) is the number of flowers insufficient to secure a satisfactory yield.

Not all flowers have an equal chance of setting fruit, however. In certain cultivars there are often high percentages of defective and staminate flowers, for reasons as yet poorly understood. The type of inflorescence and the position of individual flowers also affect fruit set. Leafy inflorescences have better chances for fruit set than purely generative, leafless inflorescences (Sauer, 1951; Jahn, 1973; Moss *et al.*, 1972). Most of the fruit set on leafless inflorescences drop and the crop is eventually borne on leafy inflorescences (Goldschmidt & Monselise, 1978). The leaves of leafy inflorescences have been assumed to play a role in provision of photosynthate, mineral nutrients or hormones to facilitate persistence of the young fruit. Erner (1989) suggested that the better water transport capacity of leafy inflorescence shoots may be responsible for the higher rate of fruit set.

Fruit set

The term 'fruit set' is commonly used to describe the process through which the flower ovary adheres and becomes fruit. For the individual fruit it is an 'all or none' event, to persist or to drop. When the ovary population of an entire tree is considered, however, the rate of fruit set assumes a quantitative meaning. The initial rate of fruit set, as observed soon after petal fall, is reduced markedly during the fruitlet abscission period. The final rate of set is determined only when fruitlet drop comes to an end, 10 to 12 weeks after anthesis.

The percentage set expresses the ratio between the rather small, final number of fruit and the initial, very large number of flowers. Thus, seemingly small deviations in percentage set make all the difference between small, average and large yields. The percentage set reveals an interesting relationship with the number of flowers. When the number of flowers is large, the percentage set may be in the range of 0.1 to 0.5%. However, when the number of flowers is small, the percentage set gets much higher, up to 10%. Thus, the rate of set compensates, to some extent, for the limited number of flowers. This indicates that the tree is able to modify its rate of drop and adjust it to its fruit-bearing potential (Goldschmidt & Monselise, 1978). Further aspects of the self-thinning mechanism of fruitlet drop have been discussed in the section on 'Fruit abscission' in this chapter.

Fruit size

At the termination of fruitlet abscission the size of the individual fruit is still small (Figure 4.21). The following few months are devoted to fruit enlargement, which requires large amounts of photosynthate. That the availability of photosynthate is indeed limiting fruit growth is demonstrated by the dramatic effects of girdling and fruit thinning. Girdling consists of removal of a ring of bark from the trunk or scaffold branches, thereby interfering with downward phloem transport. Girdling during the fruit-enlargement phase brings up to 30% increase in fruit weight (Cohen, 1984), presumably by preventing the 'escape' of photosynthate from the

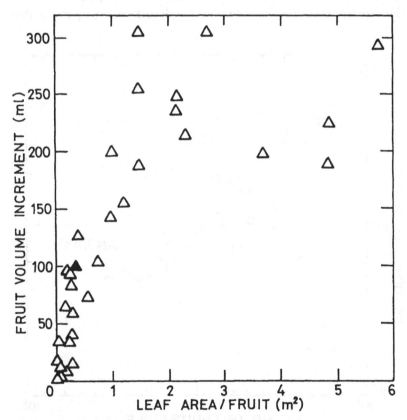

Figure 4.22 Fruit volume increment as a function of leaf area per fruit. Each point represents a single fruit of Marsh seedless grapefruit (*C. paradisi*) on partially defoliated, girdled branches. Girdling performed on May 25; the growth increment is for 32 days (June 28–July 30). The full triangle is the average growth increment of control fruit on nongirdled branches. Modified from Fishler *et al.* (1983)

girdled organ to other parts of the tree. Fruit thinning is another agrotechnique which manipulates the partitioning of photosynthate. Partial removal of fruit (= fruit thinning) increases the leaf area per fruit, thereby making more photosynthate available for each individual fruit. Figure 4.22 shows for grapefruit that fruit growth benefits from the increase in leaf area up to 1.5 m²/fruit, indicating that under most normal conditions the supply of photosynthate restricts fruit size (Fishler *et al.*, 1983).

The final yield is a product of a number of fruit by their weight at harvest. The ability of a tree to bear fruit fluctuates within a certain range, the size of fruit being inversely proportional to their number. This is true within a given cultivar, as obtained by fruit thinning, but also holds when different citrus cultivars are viewed together. Figure 4.23 shows that

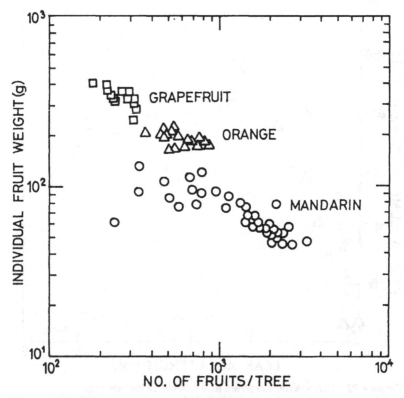

Figure 4.23 Weight of individual fruit of different Citrus species plotted against the number of fruit per tree (logarithmic scale on both axes). Data from separate experiments. Each point represents an average from all the fruit of a single tree. Modified from Goldschmidt and Monselise (1978)

mandarin, orange and grapefruit form a continuum revealing the same inverse fruit-number/fruit-size relationship. The typical fruit size of each cultivar is probably a genetic trait. Fruit number, on the other hand, seems to reflect the tree's fruit-bearing limits.

Regulation of productivity at the tree level must involve a broad array of subtle nutritional and hormonal signals. Imbalance of these regulatory systems results in productivity disorders such as alternate bearing, which is rather widespread among mandarins and mandrin hybrids (Monseliese & Goldschmidt, 1982). Yields fluctuate markedly from year to year even in regular bearing cultivars. This variability is associated primarily with climatic factors. Heat spells during the fruit set and fruitlet drop period are particularly troublesome (Reuther, 1973).

Citrus cultivars vary considerably in their yield potential for reasons that are not yet well understood. While yields of up to 120 tons per hectare are not uncommon in grapefruits, 50–70 tons per hectare would be more than satisfactory for oranges, lemons and mandarins. Actual yields are often much lower, however, due to less-than-optimal climate, light, soil, water, mineral nutrition and rootstock conditions (or because of pests and diseases). Optimization of all environmental and agrotechnical factors is required in order to maximize yields.

Recommended reading

Davenport, T. L. (1990). Citrus flowering. *Hort. Rev.*, **12**: 249–408.

Frost, H. B. and R. K. Soost. (1968). Seed reproduction, development of gametes and embryos. In *The Citrus Industry*, Vol. II, ed. W. Reuther, L. D. Batchelor and H. J. Webber, pp. 290–324. Berkeley: University of California Press.

Goldschmidt, E. E. and Monselise, S. P. (1978). Physiological assumptions toward the development of a citrus fruiting model. In *Proc. Int. Soc. Citriculture, 1977*, Vol. 2, ed. W. Grierson, pp. 668–72. Lake Alfred, Florida: ISC.

Goren, R. (1993). Anatomical, physiological and hormonal aspects of abscission in citrus. *Hort. Rev.*, **15**: 145–82.

Monselise, S. P. (1985). Citrus and related species. In *CRC Handbook of Flowering*, Vol. 2, ed. A. H. Halevy, Boca Raton, FL: CRC Press.

Monselise, S. P. (1986). Citrus. In *Handbook of Fruit Set and Development*, ed. S. P. Monselise, pp. 87–108. Boca Raton, FL: CRC Press.

Reuther, W. (1973). Climate and citrus behavior. In *The Citrus Industry*, Vol. 3, ed. W. Reuther, pp. 280–337. Berkeley: University of California Press.

Schneider, H. (1968). The anatomy of citrus. In *The Citrus Industry*, Vol. I, ed. W. Reuther, L. D. Batchelor and H. J. Webber, pp. 1–85. Berkeley: University of California Press.

Sinclair, W. B. (1984). *The Biochemistry and Physiology of the Lemon and Other Citrus Fruits.* Oakland, California: University of California, Division of Agriculture and Natural Resources. 946 pp.

Literature cited

Addicott, F. T., Lynch, R. S., Livingston, G. A. and Hunter, J. K. (1949). A method for the study of foliar abscission *in vitro. Plant Physiol.*, **24**: 537–9.

Aharoni, Y. (1968). Respiration of oranges and grapefruit harvested at different stages of development. *Plant Physiol.*, **43**: 99–102.

Bacchi, O. (1943). Cytological observations in *Citrus:* III. Megasporogenesis, fertilization and polyembryony. *Bot. Gaz.*, **105**: 221–5.

Bain, J. M. (1958). Morphological, anatomical and physiological changes in the developing fruit of the 'Valencia' orange, *Citrus sinensis* (L.) Osbeck. *Austr. J. Bot.*, **6**: 1–24.

Banerji, I. (1954). Morphological and cytological studies on *Citrus grandis* Osbeck. *Phytomorphology*, **4**: 390–6.

Bar Peled, M., Fluhr, R. and Gressel, J. (1993). Juvenile specific localization and accumulation of a rhamnosyltransferase and its bitter flavanoid in foliage, flowers and young citrus fruits. *Plant Physiol.*, **103**: 1377–84.

Bellows, T. S. Jr, Morse, J. G. and Lovatt, C. J. (1989). Modelling flower development in citrus. In *Manipulation of Fruiting*, ed. C. J. Wright, pp. 115–29. London: Butterworth & Co.

Bruck, D. K. and Walker, D. B. (1985). Cell determination during embryogenesis in *Citrus jambhiri.* I. Ontogeny of the epidermis. *Bot. Gaz.*, **146**: 188–95.

Cameron, J. W. and Frost, H. B. (1968). Genetics, breeding and nucellar embryony. In *The Citrus Industry*, Vol. II, ed. W. Reuther, L. D. Batchelor and H. J. Webber, Berkeley: University of California Press.

Cameron, J. W. and Soost, R. K. (1979). Absence of acidless progeny from crosses of acidless × acidless *Citrus* cultivars. *J. Am. Soc. Hort. Sci.*, **104**: 220–2.

Casella, D. (1935). L'agrumicultura siciliana. *Ann. R. Staz. Frutt. Agrum. Acireale N.S.*, **2**: 165–76.

Cassin, J., Bourdeaut, B., Gougue, F., Furin, V., Gaillard, J. P., Le Bourdelles, J., Montigut, C. and Monevil, C. (1969). The influence of climate upon the blooming of citrus in tropical areas. In *Proc. 1st Int. Citrus Symp.*, Vol. I, ed. H. D. Chapman, pp. 315–23. Riverside, California: University of California.

Castillo, J., Benabente, O. and Del Rio, J. A. (1992). Naringin and neohesperidine levels during development of leaves, flowers and fruits of *Citrus aurantium. Plant Physiol.*, **99**: 67–73.

Chapman, H. D. (1968). The mineral nutrition of citrus. In *The Citrus Industry*, 2nd edn., Vol. II, ed. W. Reuther, L. D. Batchelor and H. J. Webber, pp. 127–289. Berkeley: University of California Press.

Coggins, C. W. Jr, and Henning, G. L. (1988). A comprehensive California field study of the influence of preharvest applications of gibberellic acid on the rind quality of Valencia oranges. *Isr. J. Bot.*, **37**: 145–54.

Coggins, C. W. Jr and Hield, H. Z. (1968). Plant-growth regulators. In *The Citrus Industry*, Vol. II. ed. W. Reuther, L. D. Batchelor and H. J. Webber, pp. 371–89, Berkeley: University of California Press.

Coggins, C. W. and Lewis, L. N. (1962). Regreening of Valencia orange as influenced by potassium gibberellate. *Plant Physiol.*, **37**; 625–7.

Coggins, C. W. Jr, Anthony, M. F. and Fritts, R. Jr (1994). The postharvest use of GA$_3$ on lemons, in *Proc. Int. Soc. Citriculture*, 1992, Vol. I, ed. E. Tribulato, A. Gentile and G. Reforgiato, pp. 478–81. Catania, Italy: MCS Congress.

Cohen, A. (1982). Recent developments in girdling of citrus trees. In *Proc. Int. Soc. Citriculture*, 1981, Vol. I, ed. K. Matsumoto, pp. 196–9. Okitsu, Shizuoka, Japan: Okitsu Fruit Tree Research Station.

Cohen, A. (1984). Effect of girdling date on fruit size of Marsh seedless grapefruit. *J. Hort. Soc.*, **59**: 567–73.

Davenport, T. L. (1990). Citrus flowering. *Hort. Rev.*, **12**: 249–408.

Davies, F. S. (1986). Growth regulator improvement of postharvest quality. In *Fresh Citrus Fruits*, ed. W. F. Wardowski, S. Nagy and W. Grierson, pp. 79–99. Westport, CT: AVI Publishing Co.

Du Plessis, S. F. and Koen, T. J. (1988). The effect of N and K fertilization on yield and fruit size of Valencia. In *Proc. Sixth Int. Citrus Cong.*, Vol. 2, ed. R. Goren and K. Mendel, pp. 663–72. Philadelphia/Rehovot: Balaban Publishers; Weikersheim, Germany: Margraf Scientific Books.

Eaks, I. L. (1964). Ascorbic acid content of citrus during growth and development. *Bot. Gaz.*, **125**: 186–91.

Eaks, I. L. (1970). Respiratory response, ethylene production, and response to ethylene of citrus fruit during ontogeny. *Plant Physiol.*, **45**: 334–8.

Eckert, J. W. and Eaks, I. L. (1989). Postharvest disorders and diseases of citrus fruit. In *The Citrus Industry*, Vol. 5, ed. W. Reuther, E. C. Calavan and G. E. Carman, pp. 180–260. Berkeley: University of California Press.

Eilati, S. K., Monselise, S. P. and Budowski, P. (1969). Seasonal development of external color and carotenoid content in the peel of ripening 'Shamouti' oranges. *J. Am. Soc. Hort. Sci.*, **94**: 346–8.

El-Otmani, M. and Coggins, C. W. Jr (1991). Growth regulator effects on retention of quality of stored citrus fruits. *Sci. Hort.*, **45**: 261–72.

El-Otmani, M., Benismail, A., Oubahou, A. and Achouri, M. (1994). Growth regulators use on clementine mandarin to improve fruit set. In *Proc. Int. Soc. Citriculture, 1992*, Vol. I, ed. E. Tribulato, A. Gentile and G. Reforgiato, pp. 500–8. Catania, Italy: MCS Congress.

El-Otmani, M., Lovatt, C. J., Coggins, C. W. Jr and Augusti, M. (1995). Plant growth regulators in citriculture: factors regulating endogenous levels in citrus tissues. *Critical Rev. Plant Sci.*, **14**: 367–412.

Embleton, T. W., Jones, W. W. and Coggins, C. W. J. (1973a). Aggregate effects of nutrients and gibberellic acid on 'Valencia' orange crop value. *J. Am. Soc. Hort. Sci.*, **98**: 281–5.

Embleton, T. W., Reitz, H. J. and Jones, W. W. (1973b). Citrus
   fertilization. In *The Citrus Industry*, 2nd edn., Vol. 3, ed. W. Reuther,
   pp. 122–82. Berkeley: University of California Press.
Erner, Y. (1988). Effects of girdling on the differentiation of inflorescence
   types and fruit set in 'Shamouti' orange trees. *Isr. J. Bot.*, **37**: 173–80.
Erner, Y. (1989). Citrus fruit set: carbohydrate, hormone and leaf
   mineral relationships. In *Manipulation of Fruiting*, ed. C. J. Wright,
   pp. 233–42. London: Butterworth & Co.
Erner, Y., Monselise, S. P. and Goren, R. (1975). Rough fruit condition
   of the 'Shamouti' orange: occurrence and patterns of development.
   Physiol. Veg., **13**: 435–43.
Erner, Y., Goren, R. and Monselise, S. P. (1976). The induction of peel
   roughness of 'Shamouti' orange with growth regulators. *J. Am. Soc.
   Hort. Sci.*, **101**: 513–15.
Fishler, M., Goldschmidt, E. E. and Monselise, S. P. (1983). Leaf area
   and fruit size on girdled grapefruit branches. *J. Am. Soc. Hort. Sci.*, **108**:
   218–21.
Ford, E. (1942). Anatomy and histology of the Eureka lemon. *Bot. Gaz.*,
   104: 288–305.
Frost, H. B. and Soost, R. K. (1968). Seed reproduction, development of
   gametes and embryos. In *The Citrus Industry*, Vol. II, ed. W. Reuther,
   L. D. Batchelor and H. J. Webber, pp. 290–324. Berkeley: University
   of California Press.
Geraci, G., Reforgiato, G. and De Pasquale, F. (1980). Pollen tubes
   penetration into citrus styles. In *Proc. Int. Soc. Citriculture*, ed. P. R.
   Cary, pp. 58–9. Griffith, NSW, Australia.
Goldschmidt, E. E. (1976). Endogenous growth substances of citrus
   tissues. *HortScience*, **11**: 95–9.
Goldschmidt, E. E. (1988). Regulatory aspects of chloro-chromoplast
   interconversions in senescing *Citrus* fruit peel. *Isr. J. Bot.*, **47**; 123–30.
Goldschmidt, E. E. and Golomb, A. (1982). The carbohydrate balance of
   alternate-bearing citrus trees and the significance of reserves for
   flowering and fruiting. *J. Am. Soc. Hort. Sci.*, **107**: 206–8.
Goldschmidt, E. E. and Monselise, S. P. (1972). Hormonal control of
   flowering in citrus and some other woody perennials. In *Plant Growth
   Substances 1970*, ed. D. J. Carr, pp. 758–66. New York: Springer Verlag.
Goldschmidt, E. E. and Monselise, S. P. (1978). Physiological
   assumptions toward the development of a citrus fruiting model. In
   Proc. Int. Soc. Citriculture, 1977, Vol. 2, ed. W. Grierson, pp. 668–72.
   Lake Alfred, Florida: ISC.
Goldschmidt, E. E., Aschkenazi, N., Herzano, Y., Schaffer, A. A. and
   Monselise, S. P. (1985). A role for carbohydrate levels in the control of
   flowering in citrus. *Sci. Hort.*, **26**: 159–66.
Goldschmidt, E. E., Huberman, M. and Goren, R. (1993). Probing the
   role of endogenous ethylene in the degreening of citrus fruit with
   ethylene antagonists. *Plant Growth Reg.*, **12**: 325–9.
Goldschmidt, E. E., Rabber, D. and Galili, D. (1994). Fruit splitting in
   'Murcott' tangerines: control by reduced water supply. In *Proceedings
   International Society of Citriculture, 1992*, Vol. 2, ed. E. Tribulato, A.
   Gentile and G. Reforgiato, pp. 657–60. MSC Congress, Catania, Italy.
Goren, R. (1965). Hesperidin content in the Shamouti orange fruit. *Proc.
   Am. Soc. Hort. Sci.*, **86**: 280–7.

Goren, R. (1993). Anatomical, physiological and hormonal aspects of abscission in citrus. *Hort. Rev.*, **15**: 145–82.

Goren, R. and Monselise, S. P. (1964). Morphological features and changes in nitrogen content in developing Shamouti orange fruits. *Isr. J. Agri. Res.*, **14**: 65–74.

Greenberg, J., Goldschmidt, E. E. and Goren, R. (1993). Potential and limitations of the use of paclobutrazol in Citrus orchards in Israel. *Acta Hort.*, **329**: 58–61.

Grierson, W. (1986). Physiological disorders. In *Fresh Citrus Fruits*, ed. W. F. Wardowski, S. Nagy and W. Grierson, pp. 361–78. Westport, CT: AVI Publishing Co.

Grierson, W. and Ben-Yehoshua, S. (1986). Storage of citrus fruits. In *Fresh Citrus Fruits*, ed. W. F. Wardowski, S. Nagy and W. Grierson, pp. 479–507. Westport, CT: AVI Publishing Co.

Grierson, W., Cohen, E. and Kitagawa, H. (1986). Degreening. In *Fresh Citrus Fruits*, ed. W. F. Wardowski, S. Nagy and W. Grierson, pp. 253–74. Westport, CT: AVI Publishing Co.

Gross, J. (1987). *Pigments in Fruits*. London: Academic Press. 303 pp.

Guardiola, J. L. and Lazaro, E. (1987). The effect of synthetic auxins on fruit and anatomical development in 'Satsuma' mandarin. *Sci. Hort.* **31**: 119–30.

Guardiola, J. L., Garcia-Mari, F. and Agusti, M. (1984). Competition and fruit set in the Washington Navel orange. *Physiol. Plant.*, **62**: 297–302.

Haas, A. R. C. (1949). Orange fruiting in relation to the blossom opening period. *Plant Physiol.* **24**: 481–93.

Halevy, A. A. (1984). Light and autonomous induction. In *Light and Flowering Process*, ed. D. Vince-Prue, B. Thomas, and K. E. Cockshull, pp. 65–73. London: Academic Press.

Halpern, D. and Zur, A. (1988). Total soluble solids in citrus varieties harvested at various stages of ripening. In *Proc. Sixth Int. Citrus Congr.*, Vol. 4, ed. R. Goren and K. Mendel, pp. 1777–83. Philadelphia/Rehovot: Balaban Publishers; Weikersheim, Germany: Margraf Scientific Books.

Harty, A. R. and van Staden, J. (1988). The use of growth retardants in citriculture. *Israel J. Bot.*, **37**: 155–64.

Hasegawa, S. and Maier, V. P. (1982). Some aspects of citrus biochemistry and juice quality. In *Proc. Int. Soc. Citriculture, 1981*, Vol. 2, ed. K. Matsumoto, pp. 914–18. Okitsu, Shizuoka, Japan: Okitsu Fruit Tree Research Station.

Herzog, P. and Monselise, S. P. (1968). Growth and development of grapefruits in two different climatic districts of Israel. *Isr. J. Agric. Res.*, **18**: 181–6.

Hofman, P. J. (1988). Abscisic acid and gibberellins in the fruitlets and leaves of the 'Valencia' orange in relation to fruit growth and retention. In *Proc. Sixth Int. Citrus Congr.*, Vol. 1, ed. R. Goren and K. Mendel, pp. 355–62, Balaban Publishers; Philadelphia/Rehovot: Weikersheim, Germany: Margraf Scientific Books.

Holtzhausen, L. C. (1982). Creasing: formulating a hypothesis. In *Proc. Int. Soc. Citriculture, 1981*, Vol. 1, ed. K. Matsumoto, pp. 201–4, Okitsu, Shizuoka, Japan: Okitsu Fruit Tree Research Station.

Horwitz, R. M. and Gentili, B. (1986). Dihydrochalcone sweeteners from
    Citrus flavanones. In *Alternative Sweeteners*, ed. L. Nabors and R. C.
    Gerald, pp. 135–53. New York: Marcel Dekker.
Iwamasa, M. (1966). Studies on the sterility in genus *Citrus* with special
    reference to the seedlessness. *Bull. Hort. Res. Sta. Japan, Min. Agr. and
    Forest, Ser. B.*, **6**: 1–77.
Jahn, O. L. (1973). Inflorescence types and fruiting patterns in 'Hamlin'
    and 'Valencia' oranges and 'Marsh' grapefruit, *Am. J. Bot.*, **60**: 663–
    70.
Kesterson, J. W. and Hendrickson, R. (1953). The glucosides of citrus.
    Fla State Hortic. Sci., **65**: 223–6.
Kesterson, J. W., Hendrickson, R. and Braddock, R. J. (1971). Florida
    citrus oils. Florida Agricultural Experimental Station, Gainsville,
    Technical Bulletin 749, 180 pp.
Kobayashi, S., Ieda, I. and Nakatani, M. (1982). Role of the
    primordium cell in nucellar embryogenesis in *Citrus*. *Proc. Int. Soc.
    Citriculture, 1981*, Vol. I, ed. K. Matsumoto, pp. 44–8. Okitsu,
    Shizuoka, Japan: Okitsu Fruit Tree Research Station.
Koch, K. E. and Avigne, W. T. (1990). Postphloem, nonvascular
    transfer in citrus: kinetics, metabolism and sugar gradients. *Plant
    Physiol.*, **93**: 1405–16.
Koch, K. E., Lowell, C. A. and Avigne, W. T. (1986). Assimilate
    transfer through citrus juice vesicle stalks: a nonvascular portion of the
    transport path. In *Phloem Transport*, ed. J. Cronshaw, W. J. Lucas and
    R. T. Giaquinta, pp. 247–58. New York: Alan Liss.
Koltunow, A. M. (1993). Apomixis: embryo sacs and embryos formed
    without meiosis or fertilization in ovules. *Plant Cell*, **5**: 1425–37.
Krezdorn, A. H. (1982). Fruit setting of Citrus. In *Proc. Int. Soc.
    Citriculture*, Vol. 1, ed. K. Matsumoto, pp. 249–53, Okitsu, Shizuoka,
    Japan: Okitsu Fruit Tree Research Station.
Kuraoka, T. (1962). Histological studies on the fruit development of the
    satsuma orange with special reference to peel-puffing. *Mem. Ehime
    Univ. Agric.*, **8**: 105–54.
Kuraoka, T. and Kikuki, T. (1961). Morphological studies of the
    development of citrus fruit. *J. Japan. Soc. Hort. Sci.*, **30**: 189–96.
Lang, A. (1965). Physiology of flower initiation. In *Handbuch der
    Pflanzenphysiologie* XV/1, ed. W. Ruhland, pp. 1380–536. Berlin:
    Springer Verlag.
Lenz, F. (1969). Effects of daylength and temperature on the vegetative
    and reproductive growth of 'Washington' navel orange. In *Proc. 1st Int.
    Citrus Symp. Riverside*, Vol. 1, ed. H. D. Chapman, pp. 333–8.
    Riverside, California: University of California.
Lima, J. E. O. and Davies, F. S. (1984). Secondary-fruit ontogeny in
    navel orange. *Am. J. Bot.*, **71**: 532–41.
Lima, J. E. O., Davies, F. S. and Krezdorn, A. H. (1980). Factors
    associated with excessive fruit drop of navel orange. *J. Am. Soc. Hort.
    Sci.*, **105**: 502–6.
Lomas, J. and Burd, P. (1983). Prediction of the commencement and
    duration of the flowering period of citrus. *Agric. Meteorol.*, **28**: 387–96.
Lord, E. M. and Eckard, K. J. (1985). Shoot development in *Citrus
    sinensis* L. (Washington navel orange). I. Floral and inflorescence
    ontogeny. *Bot. Gaz.*, **146**: 320–6.

Lord, E. M. and Eckard, K. J. (1987). Shoot development in *Citrus sinensis* L. (Washington navel orange). II. Alteration of developmental fate of flowering shoots after GA$_3$ treatment. *Bot. Gaz.*, **148**: 17–22.

Lovatt, C. J., Streeter, S. M., Minter, T. C., O'Connell, N. V., Flaherty, D. L., Freeman, M. W. and Goodell, P. B. (1987). Phenology of flowering in *Citrus sinensis* [L.] Osbeck, cv. Washington navel orange. In *Proc. Int. Soc. Citriculture, 1984*, Vol. I, ed. H. W. S. Montenegro and C. S. Moreira, pp. 186–90. Sao Paulo, Brazil: Inst. Econ. Agricola, Centro Estadual.

Lovatt, C. J., Zheng, Y. and Kaje, K. D. (1988). Demonstration of a change in nitrogen metabolism influencing flower initiation in citrus. *Isr. J. Bot.*, **37**: 181–8.

Maier, V. P. and Beverly, G. D. (1968). Limonin monolactone, the nonbitter precursor responsible for delayed bitterness in certain citrus juices. *J. Food Sci.*, **33**: 488–92.

Maier, V. P., Hasegawa, S., Bennett, R. D. and Echols, L. C. (1980). Limonin and limonoids: chemistry, biochemistry and juice bitterness. In *Citrus Nutrition and Quality. ACS Symposium Series* 143, ed. S. Nagy and S. A. Attaway, pp. 63–82.

Marsh, A. W. (1973). Irrigation. In *The Citrus Industry*, 2nd ed., Vol. 3, ed. W. Reuther, pp. 230–79, Berkeley: University of California Press.

Monselise, S. P. (1978). Citrus fruit development: endogenous systems and external regulation. *Proceedings International Society of Citriculture, 1977*, Vol. 2, ed. W. Grierson, pp. 664–8. Lake Alfred, Florida: ISC.

Monselise, S. P. (1979). The use of growth regulators in citriculture: a review. *Sci. Hort.*, **11**: 151–62.

Monselise, S. P. (1985). Citrus and related species. In *CRC Handbook of Flowering*, Vol. 2, ed. A. H. Halevy, pp. 275–94. Boca Raton, FL: CRC Press.

Monselise, S. P. (1986). Citrus. In *Handbook of Fruit Set and Development*, ed. S. P. Monselise, pp. 87–108. Boca Raton, FL: CRC Press.

Monselise, S. P. and Goldschmidt, E. E. (1982). Alternate bearing in fruit trees. *Hort. Rev.*, **4**: 128–73.

Monselise, S. P. and Halevy, A. A. (1964). Chemical inhibition and promotion of citrus flower bud induction. *Proc. Am. Soc. Hort. Sci.*, **84**: 141–6.

Monselise, S. P. and Turrell, F. M. (1959). Tenderness, climate and citrus fruit. *Science*, **129**: 639–40.

Monselise, S. P., Goren, R. and Wallerstein, I. (1972). Girdling effect on orange fruit set and young fruit abscission. *HortScience*, **7**: 514–15.

Monselise, S. P., Weiser, M., Shafir, N., Goren, R. and Goldschmidt, E. E. (1976). Creasing of orange peel: physiology and control. *J. Hort. Sci.*, **51**: 341–51.

Moore, G. A. (1985). Factors affecting *in vitro* embryogenesis from undeveloped ovules of mature *Citrus* fruit. *J. Am. Soc. Hort. Sci.*, **110**: 66–70.

Moss, G. I. (1969). Influence of temperature and photoperiod on flower induction and inflorescence development in sweet orange (*Citrus sinensis* L. Osbeck). *J. Hort. Sci.*, **44**: 311–20.

Moss, G. I. (1970). Chemical control of flower development in sweet orange (*Citrus sinensis*). *Austr. J. Agric. Res.*, **21**: 233–42.

Moss, G. I. (1977). Major factors influencing flower formation and subsequent fruit-set of sweet orange. In *Primera Congreso Mundial de Citricultura, 1973*, Vol. 2, ed. O. Carpena, pp. 215–22. Murcia, Valencia, Spain: Ministerio de Agricultura, Instituto de Investigaciones Agrarias.

Moss, G. I., Steer, B. T. and Kriedemann, P. E. (1972). The regulatory role of inflorescence leaves in fruit setting by sweet orange (*Citrus sinensis*). *Physiol. Plant*, **27**: 432–8.

Newman, J. E., Cooper, W. C., Reuther, W., Cahoon, G. A. and Peynado, A. (1967). Orange fruit maturity and net heat accumulation. In *Ground Level Climatology*, ed. R. H. Shaw, pp. 127–47. American Association for the Advancement of Science Publication No. 86.

Nitsch, J. P. (1953). The physiology of fruit growth. *Ann. Rev. Plant Physiol*, **4**: 199–236.

Osawa, L. (1912). Cytological and experimental studies in *Citrus. J. Coll. Agr. Univ. Tokyo*, **4**: 83–116.

Osborne, D. J. (1989). Abscission. *CRC Crit. Rev. Plant Sci.*, **8**: 103–29.

Pharis, R. P. and King, R. W. (1985). Gibberellins and reproductive development in seed plants. *Ann. Rev. Plant Physiol.*, **36**: 517–68.

Randhawa, G. S., Nath, N. and Choudhury, S. S. (1961). Flowering and pollination studies in citrus with special reference to lemon (*Citrus limon* Burn.). *Indian J. Hort.*, **18**: 135–47.

Reuther, W. (1973). Climate and citrus behavior. In *The Citrus Industry*, 2nd ed., Vol. 3, ed. W. Reuther, pp. 280–337, Berkeley: University of California Press.

Saks, Y., Weiss, B., Chalutz, E., Livne, A. and Gepstein, S. (1988). Regreening of stored pummelo fruit. In *Proc. Sixth Int. Citrus Congr.*, Vol. 2, ed. R. Goren and K. Mendel, pp. 1401–6. Philadelphia/Rehovot: Balaban Publishers; Weikersheim, Germany: Margraf Scientific Books.

Sasson, A. and Monselise, S. P. (1977). Organic acid composition of 'Shamouti' oranges at harvest and during prolonged postharvest storage. *J. Am. Soc. Hort. Sci.*, **102**, 331–6.

Schaffer, A. A., Goldschmidt, E. E., Goren, R. and Galili, D. (1985). Fruit set and carbohydrate status in alternate and nonalternate bearing *Citrus* cultivars. *J. Am. Soc. Hort. Sci.*, **10**: 574–8.

Schneider, H. (1968). The anatomy of citrus. In *The Citrus Industry*, 2nd edn., Vol. II, ed. W. Reuther, L. D. Batchelor and H. J. Webber, pp. 1–85, Berkeley: University of California Press.

Sinclair, W. B. (1984). *The Biochemistry and Physiology of the Lemon and Other Citrus Fruits*. University of California, Division of Agriculture and Natural Resources, 946 pp.

Soule, J. and Grierson, W. (1986). Maturing and grade standards. In *French Citrus Fruits*, ed. W. F. Wardowski, S. Nagy and W. Grierson, pp. 23–48. Westport, CT: AVI Publishing Co.

Southwick, S. M. and Davenport, T. L. (1986). Characterization of water stress and low temperature effects on flower induction in citrus. *Plant Physiol.*, **81**: 26–9.

Tatum, J. H., Hearn, C. J. and Berry, R. E. (1978). Characterization of *Citrus* cultivars by chemical differentiation. *J. Am. Soc. Hort. Sci.*, **103**: 492–6.

Theologis, A., Zarembinski, T. I., Oeller, P. W., Liang, X. and Abel, S. (1992). Modification of fruit ripening by suppressing gene expression. *Plant Physiol.*, **100**: 549–51.

Thomson, W. W., Lewis, L. N. and Coggins, C. W. (1967). The reversion of chromoplasts to chloroplasts in Valencia oranges. *Cytologia*, **32**: 117–24.

Ting, S. V. and Attaway, J. A. (1971). Citrus fruits. In *The Biochemistry of Fruits and Their Products*, Vol. 2, ed. A. C. Hulme, pp. 107–69. London: Academic Press.

Ting, S. V. and Rouseff, R. L. (1986). *Citrus Fruits and Their Products: Analysis and Technology*. New York: Marcel Dekker. 293 pp.

Tzur, A. (1994). Carbohydrate metanolism in various stages of citrus fruit development. PhD dissertation, The Hebrew University of Jerusalem, 165 pp. (In Hebrew with English summary.)

Ulrich, R. (1970). Organic acids. In *The Biochemistry of Fruits and Their Products*, Vol. 1, ed. A. C. Hulme, pp. 89–114 London: Academic Press.

Vardi, A., Frydman-Shani, A. and Weinbaum, S. A. (1988). Assessment of parthenocarpic tendency in Citrus using irradiated marker pollen. In *Proc. Sixth Int. Citrus Congr.*, Vol. I, ed. R. Goren and K. Mendel, pp. 225–30. Philadelphia/Rehovot: Balaban Publishers; Weikersheim, Germany: Margraf Scientific Books.

Wakana, A. and Uemoto, S. (1987). Adventive embryogenesis in Citrus (Rutaceae). I. The occurence of adventive embryos without pollination and fertilization. *Am. J. Bot.*, **74**: 517–30.

Wakana, A. and Uemoto, S. (1988). Adventive embryogenesis in Citrus (Rutaceae). II. Post fertilization development. *Am. J. Bot.*, **75**: 1031–47.

Wallace, A., Abou-Zamzam, A. M. and Procopiou, T. (1977). Organic acid synthesis in sour and sweet lemon fruits. In *Primo Congreso Mundial de Citricultura, 1973*, Vol. 2, ed. O. Carpena, pp. 305–10, Murcia, Valencia, Spain: Ministerio de Agricultura, Instituto Nacional de Investigaciones Agrarias.

Wilson, W. C., Coppock, G. E. and Attaway, J. A. (1982). Growth regulators facilitate harvesting of orange. In *Proc. Int. Soc. Citriculture, 1981*, Vol. I, ed. K. Matsumoto, pp. 278–81. Okitsu, Shizuoka, Japan: Okitsu Fruit Tree Research Station.

Yokoyama, H. and Keithley, J. H. (1991). Regulation of biosynthesis of carotenoids. In *Plant Biochemical Regulators*, ed. H. W. Gausmann, pp. 19–25. New York: Marcel Dekker.

Young, L. B. and Erickson, L. C. (1961). Influence of temperature on color change in Valencia oranges. *Proceedings of the American Society of Horticultural Science*, **78**: 197–200.

Zacharia, D. B. (1951). Flowering and fruit setting of Shamouti orange trees. *Palest. J. Bot.* (Rehovot Ser.), **8**: 84–94.

5

Aspects of cultivated citrus

Orchard design and spacing

WITH CAREFUL ADJUSTMENT of rootstocks and cultural practices citrus can be grown satisfactorily on a wide range of soils. In general, the deep, well-drained sandy loam soils are best suited for citrus production. No single characteristic of good citrus soil is more essential than good drainage. Without satisfactory drainage, accumulation of free water in the root zone results in poor aeration and injury to roots. In regions of heavy rainfall, the use of shallow soils with impervious subsoils or with hardpan is particularly hazardous, because under such conditions roots are most susceptible to fungal infection. Lack of drainage also contributes to effects caused by salinity which, in turn, may reduce yields. As salinity of irrigation water increases, it is necessary to move more water through the root zone to carry out accumulated salts. Thus, any restriction of drainage becomes especially harmful where irrigation is practiced.

In detailed planning of the orchard, decisions regarding planting distances and tree spacing are most important. For many years, considerations have centered on the distance which would optimize yields during the entire life of the orchard (several decades), also allowing sufficient space to conduct necessary cultural operations. Trees have been planted at distances assuring adequate light for the tree and the passage of equipment at maturity. Thus, distances between trees (varying, of course, with variety, rootstock, etc.), have generally ranged 6 to 10.5 meters in either direction, resulting in tree densities of 86–270 trees per hectare in California, Florida and South Africa.

Most recently, increased cost of land, of irrigation water and of cultural practices have prompted growers in many areas of the world to seek maximum early production to compensate for higher costs. Consequently, the trend has been to reduce distances between trees and provide a greater number of trees per hectare. The high cost of manual labor required for harvesting large-sized trees has further favored this trend. On the other extreme, many old orchards in the Mediterranean area used to have high tree densities, ranging up to 1000 and, in some cases 1500, trees per hectare. Under such crowded conditions, most of the canopy was shaded out and fruit was produced only in the upper portion of the tree, where light conditions are satisfactory, leading to overall low production.

Modern trends in the culture of deciduous trees have further contributed toward the development of the concept of the compact, highly productive citrus tree. Thus, the control of tree size has become a major issue in orchard management. Ideally, appropriate stock/scion combinations should provide for the desirable tree size. Such combinations are available in citrus only in a few cases. Considerable efforts have been made, therefore, in the search for dwarfing rootstocks. Inoculation with mild strains of the exocortis viroid has been shown to control effectively tree size in Marsh grapefruit and other varieties (see Diseases, Chapter 5). Maintenance of tree size and adequate light may be accomplished by pruning, which is usually carried out mechanically through 'hedging' and 'topping'. These are rather costly operations, however, which involve the removal of peripheral fruit layers of the canopy and which are therefore usually undertaken only once every two to three years. Root restriction is another means of controlling tree size. The drip irrigation technique, which involves concentration of the root system mostly in the soil volume surrounding the dripper, can be utilized to restrict tree size. Growth retardants such as paclobutrazol have also been tried as soil drenches, for the same purpose.

Rootstocks

Easy propagation by seed and transport of seed facilitated expansion of citrus to new environments. Apomixis also allowed propagation true to type of worthy genotypes. Citrus trees were generally grown from seed till the mid 1800s. They are still grown from seed in certain areas of Central and South America and in South East Asia. In the latter environment pummeloes and mandarins are also propagated by air layers (marcottage). As a result of grave damage to seedling trees by *Phytophthora* foot rot, in the Azores (around 1842) and elsewhere, with orange seedlings being particularly sensitive, the transition to the use of budded trees in citriculture began in most environments. Seedling citrus trees also show a high degree of juvenility, which is often associated with later bearing. Citrus cultivars (scions) are budded or grafted onto highly apomictic rootstocks propagated from seed. Trees are thus composed of two components: scion and rootstock. The horticultural performance of a citrus tree is the result of the reciprocal interaction between the tree's genetic components.

The rootstock has a large effect on scion vigor and size, fruit size, yield, fruit and juice quality as well as on tolerance to salt, cold and drought. Rootstocks differ widely in tolerance to *Phytophthora*, viruses and nematodes. Citrus rootstocks also have a considerable effect on leaf mineral content in the scion (Wutscher, 1989). The main characteristics of the leading rootstocks are given in Table 5.1.

Rootstocks have also to be acceptable nursery plants and to show good and prolonged compatibility with scion varieties. Compatibility has been assessed by bud union smoothness, absence of anatomical and morphological abnormalities and satisfactory tree vigor. An adequate number of seeds in the fruit and a high degree of nucellar embryony are required to facilitate commercial propagation of rootstocks by seed.

Rootstocks have contributed to a very large extent to successes and failures in citrus industries. At present, smaller, high-yielding trees are sought for closer plantings. However, in contrast to the apple, no satisfactory dwarfing stocks in citrus are as yet generally available. Certain rootstock combinations contribute to high yields per unit of land at close spacings. Efforts to reduce citrus tree size include experiments with tetraploid rootstocks, trials of the semi-dwarfing Rangpur × Troyer (sensitive to *Phytophthora* and citrus blight) and of the dwarfing rootstock 'Flying Dragon'. The latter, probably a mutant of *Poncirus trifoliata*, reduced canopy volume of 9-year-old Valencia trees to a third of that of trees on standard rootstocks (Roose *et al.*, 1994).

Citrus irrigation and water use

Irrigation is the most costly practice in citrus growing in arid and semi-arid climates with long dry periods. It is also increasingly employed in humid and sub-humid climates to maintain yields that are often reduced because of dry, rainless periods. About 66% of citrus groves in Florida have been reported to receive at least supplemental irrigation. Koo (1979) found Valencia orange responding by an average of a 22% increase of yield in eight years out of nine, with 27.6 cm $ha^{-1}y^{-1}$ irrigation added to 116 cm $ha^{-1}y^{-1}$ rain.

Water use involves loss through transpiration from the crop and evaporation from the soil. The sum of the two components of water loss is termed evapotranspiration (ET). ET will be a function of the stage of

Table 5.1 *Characteristics of the principal citrus rootstocks*

Common name	Botanical classification	Main characteristics
Sour orange	*Citrus aurantium*	Cold tolerance; foot rot tolerance; good compatibility; vigorous; high fruit quality; highly sensitive to tristeza virus; often relatively low early yields
Rough lemon	*Citrus jambhiri*	Deep rooted large trees; high susceptibility to *Phytophthora*, blight, low fruit quality; high early yields
Volkamer lemon	*Citrus volkameriana*	Similar to rough lemon; cold hardier, more tolerant to *Phytophthora parasitica*
Rangpur lime	*Citrus reticulata* var *austera* × *Citrus limon*	High early yields, salt tolerance; fruit quality mediocre; sensitive to *Phytophthora*
Alemow	*Citrus macrophylla*	High early yields; susceptible to tristeza, xyloporosis; fruit quality moderate to low; sensitive to cold, blight
Sweet orange	*Citrus sinensis*	Tolerant to tristeza; fruit quality high; very susceptible to *Phytophthora*; moderate to high blight tolerance
Cleopatra mandarin	*Citrus reticulata*	Small fruit size; salt tolerance; cold tolerance; fruit quality high; slow growth in nursery; relatively blight tolerant; tolerant of high pH
Trifoliate orange	*Poncirus trifoliata*	Smaller than standard tree; large, high quality fruit; high tolerance to tristeza, *Phytophthora*; low tolerance to salt and high pH; highly susceptible to exocortis, drought
Carrizo citrange	*Citrus sinensis* × *Poncirus trifoliata*	High yield and good fruit quality; susceptibility to exocortis; tolerant to burrowing nematode
Troyer citrange	*Citrus sinensis* × *Poncirus trifoliata*	Similar to Carrizo; no resistance to burrowing nematode; relatively good tolerance to cold
Swingle citrumelo	*Citrus paradisi* × *Poncirus trifoliata*	Compatibility often inadequate; fruit quality medium to high, sensitive to calcareous soils; comparatively cold hardy; vigorous; high early yields with grapefruit; good tolerance to blight

plant growth, the crop and the evaporative demand of the atmosphere, depending on soil water availability. If the crop is subject to stress, ET will decline. During establishment of the orchard, ET will be lower than that of mature, bearing trees. Increase in crop load will generally increase ET. Citrus water use efficiency (WUE), representing the amount of CO_2 fixed per amount of water transpired, is rather low compared with that of many other crop plants. Basic aspects of citrus water relations have been extensively discussed by Kriedemann and Barrs (1981) and by Jones et al. (1985). Citrus tree irrigation, including various aspects of irrigation and salinity, has been reviewed by Shalhevet and Levy (1990).

As early as 1646, Ferrarius noted that citrus trees require abundant moisture but cannot endure stagnant water. Citrus roots are only slightly less sensitive to deficient aeration than avocado roots. They seem to be highly sensitive to hydrogen sulfide, which arises in flooded soils due to the activities of sulfur-fixing bacteria.

While *Citrus* is considered a typical mesophyte, the leaves have many xeromorphic characteristics. The adaxial epidermis is covered by a thick, waxy cuticle. The rigid leaf shows wilting only at low leaf water potentials. Osmotic potential rarely rises over -10 bars, even in a relatively turgid leaf. Under favorable conditions, high transpiration rates are observed, while transpiration is limited under unfavorable edaphic and atmospheric moisture conditions. Under high evaporation demand stomata generally close. Stomata are much more abundant on the lower leaf surface than on the adaxial surface (800 mm^{-2} versus 40 mm^{-2}). The high stomatal density predisposes citrus toward potentially high transpiration, but the network of first and second order leaf veins is relatively sparse. Leaf life is usually between nine and 24 months. The largest leaf drop occurs immediately after blossoming. Hilgeman et al. (1969) stated that apparent transpiration and water stress of orange trees in summer was similar in Arizona and Florida. Evaporative demand was much higher in the desertic environment of Arizona.

Girton (1927) found the minimum temperature for root development to be 12 °C, the optimum to be 26 °C, and maximum to be 37 °C. The decreased water absorption at low root temperatures is due mainly to a decrease in permeability of root membranes and an increase in the viscosity of water (Figure 5.1). Feeder roots are sparse in *Poncirus*, and rather abundant in Rough lemon.

Castle (1980) noted that the shallow suberized root system was generally equipped with only vestigial root hairs. Root hydraulic conductivity in citrus is relatively low. Hydraulic conductivity increases with higher root temperature (Syvertsen, 1981).

Effect of irrigation on vegetative growth, fruit quality and various aspects of orchard management

Water stress significantly limits canopy development. During the first 20 years in an orchard, relationship between canopy volume and yield was found to be positive (Levy *et al.*, 1978). Cases of excessive growth in full-sized trees may lead to decreased yield, due to shading and other factors. While vegetative growth is a continuing process, fruit production is a result of the reproductive process, followed by phases of fruit set and development.

Moderate water stress seems often to favor reproductive development, while fruit set and enlargement depend upon turgor relations (Kriedemann and Barrs, 1981). Yield is thus a function of both amount and timing of irrigation (in addition to rainfall). Water deficit may often substitute for the cold requirement in promoting flowering. It may also shift the balance from excessive vegetative growth towards reproductive growth, and increase cold hardiness (Yelenosky, 1979). Severe water stress will inhibit vegetative and fruit growth.

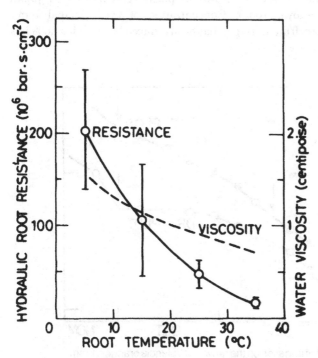

Figure 5.1   Effect of temperature on flow of water through lemon roots. After Ramos and Kaufmann (1979)

Overirrigation, especially surface irrigation, may wet the trunks of trees and increase the incidence of foot rot caused by *Phytophthora*. Lime-induced chlorosis has been aggravated by overirrigation, and tends to be reduced by drip irrigation. Irrigation timing is considered crucial for reproductive development, fruit set and fruit enlargement. However, cropping in one season influences both root extension and top growth, often with a carry-over effect on yield in the successive year.

A lowering of yield may precede any noticeable decrease of tree vigor. Increased irrigation generally leads to an increase in size of individual fruits (Figure 5.2). A highly complicating factor is crop size, affecting competition between fruits for photosynthates and growth substances. Fruit rind color seems to decrease somewhat with increased irrigation (Bielorai *et al.*, 1982). Acid concentration and the ratio between the amount of sugar in the juice (total soluble solids, TSS) and acid are important in defining fruit quality, fruit value and the timing of picking. Water shortage often causes an increased TSS in the juice; however, acidity may increase even more. Irrigation will generally increase juice content. Moisture supply is cut off at the ripening stage to increase TSS of juice in Satsuma mandarins grown in plastic greenhouses in Japan (Mukai and Kadoya, 1994). Concentration of the principal bitter compound in grapefruit, naringin, has been reduced by applying 50% less

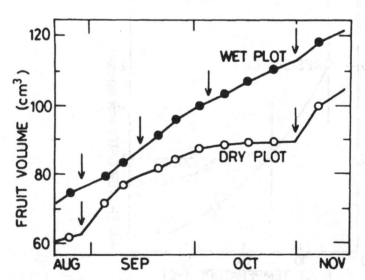

Figure 5.2   Effect of irrigation on the growth of Valencia oranges. From Hilgeman and Van Horn (1954). Courtesy of Arizona Agricultural Experiment Station

water per irrigation (6 cm-ha^{-1} compared with 12 cm-ha^{-1}) (Cruse *et al.*, 1982).

Scheduling irrigations

The total seasonal amount of irrigation water needed by a mature orchard for optimum yields depends on the daily course of evapotranspiration, rainfall distribution, the stionic (scion/stock) combination, water quality and soil factors.

In many cases a simple calendar schedule of irrigation, without regard to seasonal requirements, is being followed, mainly because of water rights. However, less water may be required in the spring and fall than during summer. On the other hand, spring applications may often prove crucial because of hot spells during fruit set and the 'June drop' period.

Of the physiological indicators proposed, fruit growth seems to be of greatest interest (Furr and Taylor, 1939). It applies only to fruiting trees and, moreover, fruit growth does not proceed uniformly during the season. Fruit volume increase was found to show correlation with stomatal aperture (Oppenheimer and Elze, 1941), leaf water potential (Ashizawa *et al.*, 1981), and, most significantly, with soil suction (Lombard *et al.*, 1965).

Further physiological indicators tested, mainly for research purposes, include leaf water potential (Kaufmann, 1968), intact leaf moisture level (Peynado and Young, 1968), trunk growth (Hilgeman, 1963) and sap velocity (Cohen *et al.*, 1981). Determining irrigation need by use of tensiometer instruments is practiced by many growers and researchers. Tensiometer measurements register the soil water potential, a function of soil water content in the root zone at the location placed. When correlated with tree performance (including fruit growth), tensiometer measurements serve as a guide for determining the timing of irrigation (Marsh, 1973). Placing tensiometers in pairs at two depths is common practice. Results are expressed as soil suction (in centibars). Figure 5.3 shows the correlations between soil suction values and fruit growth, measured by fruit circumference. The use of the neutron scattering method to measure soil water content has been limited essentially to research purposes.

Another method employed for determining irrigation timing and amount is the measurement of Class A pan evaporation, and conversion to ET through the use of an empirical crop coefficient. Van Bavel *et al.* (1967) estimated a crop coefficient value of 0.66 in fruit-bearing orange orchards. Shalhevet and Bielorai (1978) gave an estimate of 0.6 for grapefruit for the ratio of ET to Class A pan evaporation for optimum yield.

The ratio between evapotranspiration (ET) and evaporation from a free water surface (E_0) has been defined by Penman (1948) as the crop factor (f). The relationship $ET = fE_0$ then serves as a base for irrigation scheduling. E_0 can be derived either from meteorological data or empirically by measuring evaporation from a Class A pan (E_p). Crop factor (f) was rated as 0.68 in Israel and 0.93 in Arizona. ET values of 85 ha-cm y^{-1} for orange have been reported, compared with values of E_0 of 157 ha-cm y^{-1}, with most of the difference attributed to canopy resistance (Stanhill, 1972). Some use the simpler pan coefficient, K_p. K_p values for the orchards in Israel and Arizona were 0.54 and 0.66 respectively, as $E_0 = 0.8 E_p$ (Kriedemann and Barrs, 1981). The crop factor will vary throughout the season. It seems that citrus orchards transpire less water per unit land surface than many agricultural crops. This can be perhaps attributed to the high gaseous diffusive resistance in leaves, possibly coupled with the relatively low hydraulic conductivity in the tree's vascular system.

Effect of irrigation methods

Irrigation methods in citrus have been reviewed by Marsh (1973) and by Shalhevet and Levy (1990). Methods include gravity irrigation, which

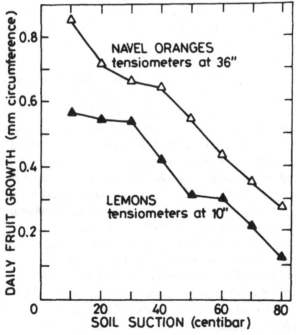

Figure 5.3　Effect of soil suction on fruit growth rate. After Beutel (1964)

requires larger water quantities, various systems of sprinkler irrigation, and more recently, microirrigation (including minisprinklers, sprayers and drip irrigation, also known as trickle irrigation).

Drip irrigation is a method by which water is applied from emitters spaced on a distribution line placed along the tree row, with usual rates of application, ranging from 2 to 8 l h^{-1} per emitter. Microsprinkler emitters deliver usually 20 to 80 l h^{-1}. As a more limited root volume is wetted, the need for frequent irrigation arises. The limited volume of wetted soil restricts root growth and constrains tree volume. The system can be also conducive to the reproductive process and to cropping on small trees when 'fertigation' is used (see below). Microirrigation systems require more intensive management and measures to overcome eventual plugging of the system. Their use enables better control of soil aeration.

With limited water supply and lower water quality, much attention is turned to water use efficiency as well as to the development of more sophisticated forms of water management and distribution.

Use of reclaimed wastewater for irrigation is on the increase, in view of the competition for water with urban areas and industry. Treated wastewater, which is relatively inexpensive, contains various amounts of mineral elements required as fertilizers. Some sources may be dangerously high in salt, boron or metals.

Irrigation is often coupled with fertilizer application ('fertigation') and less frequently with herbicide application ('herbigation'). Drip-irrigated orchards nearly always use fertigation. Koo (1984) reported on fertigation in Florida, stating that coverage of 40% of the ground is necessary in order to obtain results comparable with those obtained by applying dry fertilizers. Under-canopy sprinklers have also been successfully used for the application of herbicides and certain fungicides.

Use of citrus irrigation for micro-climate control includes attempts at cold hardening and frost protection, and reduction of flower and fruitlet drop by evaporative cooling.

With radiation freeze and a strong temperature inversion, use of microsprinklers resulted in one case in a 6–8 °C increase in temperature. Overhead sprinkling may cause ice accumulation on the foliage, causing limb breakage (Parsons et al., 1986).

Periods of drought, followed by rainfall or irrigation often induce profuse bloom. Out-of-season flowering can thus be induced, as practiced in Sicily with lemons, resulting in a significant summer crop of lemons (verdelli).

Also, the crucial role of a favorable tree water balance shortly after bloom on increased fruit set in Navel orange has been repeatedly observed. Water stress during flowering and 'June drop' may have serious deleterious effects on yield. Modification of microclimate by irrigation is also practiced for protection against heat.

FOLIAR DAMAGE THROUGH SPRINKLER SPRAY

Harding *et al.* (1956) first described severe damage to leaves in the skirt of citrus trees of under-tree sprinkler irrigated orchards, even when the irrigation used water of good quality (2–3.3 Mm^{-3} Cl^-). Citrus leaves accumulate Cl^- and Na^+ from direct contact with water drops. Accumulation is a function of evaporation rate, resulting in increased salt concentration in the water film on the leaves. Washington Navel orange accumulates salts faster through the leaves than does the less salt tolerant avocado. Changes in overhead sprinkler irrigation from daytime to nighttime, or changing to under-tree irrigation significantly reduced foliar concentration of Cl^-.

Salinity

Salinity problems develop most often in arid and semi-arid regions when the amount of water applied is insufficient to result in adequate leaching. Citrus ranks among salt-sensitive plants (Bielorai *et al.*, 1988; Alva and Syvertsen, 1991). As early as 1900 (Loughridge) it was noted that the various citrus species are extremely sensitive to chlorides. NaCl is an important constituent in many soils, and in irrigation water in various semi-arid regions where citrus is grown. It seems that when the dominant anion in the soil solution is Cl^- and the rootstock readily absorbs this anion its toxicity may overshadow the total salt effect. Response of Marsh grapefruit on sour orange rootstock was postulated by Bielorai *et al.* (1978) to be dominated by the total salt (osmotic) effect, chloride being the dominant anion in the irrigation water, with relatively low leaf accumulation of Cl^-.

Citrus trees injured by sodium salts are highly susceptible to adverse climatic conditions. Visible symptoms of salt injury (chloride effect) show tip yellowing followed by tip burn. Progressive yellowing and necrosis proceed downward.

Mechanism

Salt tolerance in a range of crop plants is associated with salt exclusion, the ability to restrict and uptake and/or transport salt between roots and shoots (Levitt, 1980). Rangpur lime was shown to exclude Cl^- from shoots, not by sequestering it in roots, but by restricting its entry into and

Table 5.2 *Cl⁻, Na⁺ and K⁺ concentrations in leaves, stems and roots of Rangpur lime, sweet orange and* Poncirus trifoliata *plants treated with 50 mM NaCl for six weeks*

Plant part	Rootstock	Cl^-	Na^+	K^+
Leaves	Rangpur lime	50±4	60±5	192±6
	Sweet orange	79±7	88±14	146±8
	Trifoliata	140±8	25±4	320±12
Stems	Rangpur lime	84±3	121±7	89±3
	Sweet orange	96±5	83±2	134±4
	Trifoliata	76±2	55±8	166±16
Roots	Rangpur lime	89±2	96±9	107±4
	Sweet orange	98±4	97±3	147±8
	Trifoliata	106±7	109±7	138±5

From Grieve and Walker (1983).

within the roots. Root membranes seem implicated (Walker and Douglas, 1983). The regulation of Na^+ distribution in *Poncirus trifoliata* has been shown to be independent of the mechanism regulating Cl^- distribution (Table 5.2). Na^+ seems to accumulate preferentially in basal segments (Grieve and Walker, 1983).

Genetic component resistance of rootstocks to salinity

Considerable genetic variability in salt uptake and tolerance has been noted in citrus, especially between rootstocks. With few exceptions, the documented effects of salinization in citrus are based on short-term responses of ungrafted rootstocks.

Table 5.3 gives the relative tolerance of a comparatively large number of genotypes to salt, boron and lime. It seems that the Cl^- and Na^+ accumulating properties of a particular species may be quite different. Thus, Rangpur lime and Cleopatra mandarin are proven Cl^- excluders, while *Poncirus trifoliata* seems to exclude Na^+ (Grieve and Walker, 1983; Walker, 1986).

The very high tolerance of Rangpur lime is shown by the very low mortality of Minneola budded onto Rangpur lime (Table 5.4) and the low Cl^- in the leaves even after three years of irrigation with extremely high Cl^- irrigation water (Vardi et al., 1988). The latter combination proved much more salt tolerant than Shamouti orange on the same rootstock (Figure 5.4).

Table 5.3 *Classification of salt, boron and lime tolerance[1] of citrus varieties used in Texas as rootstock for two red grapefruit varieties*

	Salt	Boron	Lime
Cleopatra	g	m–p	m
Sunki	g	—	m
Ponkan	m	m	p
Dancy	m	—	m
Clementine	m	p	p
Willow leaf	m	—	m
Satsuma	p	—	p
King	p	—	m
Temple	—	—	p
Orlando	m	—	p
Minneola	m	m	m
Mexican lime	m	m	m
Sweet lime	p	p	m
Rangpur lime	g	m	g
Rough lemon	m	m	g
Calamondin	m	p	m
Poncirus	p	m	p
Troyer	p	g	p
Carrizo	m	—	m
Citrumelo Sacaton	m	m	p
Citrumelo CPB4475	m	m	p
Pineapple orange	m	g	p
Hamlin orange	m	—	p
Valencia	m	—	p
Sour orange	m	m	g
Duncan	m	m	p
Redblush	—	m	p
Siamese pummelo	m	—	p
Severinia buxifolia	g	g	p
Citron PI 11292	p	p	—

[1] p = poor, m = medium, g = good.

Modified from Cooper *et al.* (1956), as published by Chapman (1968).

Effect of salinity on yield

A three-year experiment in a 20-year-old orange grove showed no differences in yield in response up to 13 mM^{-3} Cl$^-$ in the irrigation water (EC$_i$ = 1.8 dSm^{-1} (decisiemens)). Tolerance of orange and grapefruit in terms of electrical conductivity (EC$_e$) of the soil saturation extract was reported by Maas and Hoffman (1977). The threshold of the salinity effect was 1.7–1.8 dSm^{-1} with a yield reduction of 16% per 1 dSm^{-1} increase in EC$_e$.

Table 5.4 *Plant mortality (percentage in brackets) at the end of a three-year experiment with Minneola and Shamouti on six rootstocks irrigated with water containing up to 3000 mg Cl⁻/l*

Rootstock	Cultivar	
	Minneola tangelo	Shamouti orange
Sour orange	11 (39%)	15 (54%)
C. volkameriana	7 (25%)	12 (43%)
Troyer citrange	6 (25%)	5 (21%)
Poncirus trifoliata	4 (15%)	6 (23%)
Swingle citrumelo	3 (12%)	4 (15%)
Rangpur lime	1 (4%)	0

From Vardi *et al.* (1988).

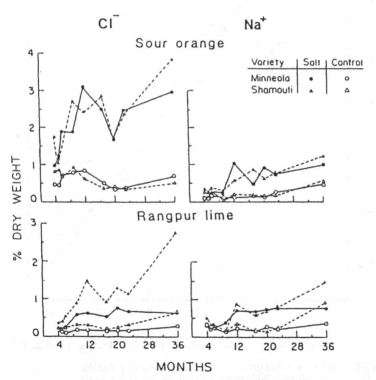

Figure 5.4 Leaf Cl⁻ and Na⁺ content of two cultivars (Minneola tangelo and Shamouti orange) budded on Rangpur lime and sour orange. Control irrigated with 250–300 mg Cl⁻ l⁻¹; salt treatment with up to 3000 mg Cl⁻ l⁻¹. Adapted from Vardi *et al.* (1988)

A further comparison of relative yield as a function of electrical conductivity of the soil saturation extract was presented by Shalhevet and Levy (1990) (Figure 5.5). A threshold salinity of 1.3 dS with a relative yield reduction of 13% per 1 dSm⁻¹ increase in EC is deducted.

Bielorai *et al.* (1978) report yield to be linearly related to mean Cl⁻ concentration in the soil saturation extract, with 1.45% yield reduction for each 1 meq/1 increase in Cl⁻ concentration above the threshold (4.5 meq/1).

4 mmho/cm in the saturation extract was associated with a 50% decrease in trunk area (Pearson *et al.*, 1957).

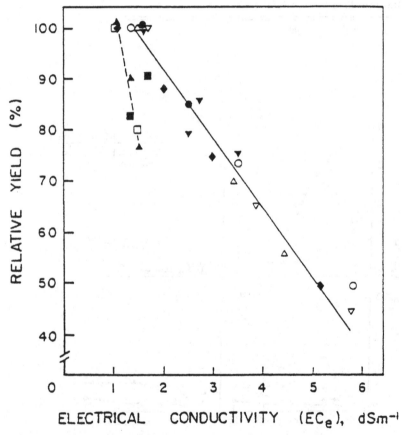

Figure 5.5 Relative yield of citrus varieties and rootstocks plotted against electrical conductivity of soil saturation extract (EC_e). Regression equation $Y/y_m = 1.0-0.129$ ($EC_e = 1.28$). From Shalhevet and Levy (1990)

Mineral nutrition

Use of mineral nutrients by citrus

Mineral nutrition of citrus has been widely studied. Recent trends in citrus irrigation, which often result in a more limited root system, have stimulated new interest in citrus mineral nutrition and fertilization.

This discussion will concern the three macronutrients, nitrogen, potassium and phosphorus, magnesium, and six micronutrients that generally influence and often limit citrus production in various environments. Further nutrients, such as calcium and sulphur, may be deficient in certain situations (Chapman, 1968). Mineral nutrients are required for various physiological processes and structural components (Table 5.5).

Nitrogen (N) concentrations in citrus tissues are highest in the leaves and are next highest in immature fruits (Table 5.6). Leaves contain enzymes for the PCR cycle, carbohydrate metabolism and nitrate reduction. Nitrogen is a structural component of chlorophyll and an important constituent of proteins. It is essential for cell division and expansion. Acute lack of shortage of this element arrests vegetative growth and results in bronzing or yellowing of foliage. New leaves of trees deficient in N are thin and fragile, and the angle between stem and leaf is rather narrow. Citrus orchards deficient in N may exhibit a decrease in flowering and fruiting even before striking leaf symptoms develop. This is largely due to a reduction in growth. Over-fertilization with N may cause excess growth, reduction of yield and decrease in fruit quality.

Large amounts of potassium (K) are required by citrus. Potassium is an important constituent of the fruit (40% of the total mineral content). It is involved in the translocation of carbohydrates. It acts as an osmotic agent in the opening and closing of stomata. It plays an important role in controlling the acidity of the fruit juice. It functions in charge balancing and membrane transport. Reduction in yields due to K deficiency is attributed mainly to reduced vegetative growth. However, the effect of low K on decreasing average fruit size is striking.

Phosphorus (P) is a highly mobile element within the tree. Phosphorus is a component of nucleoproteins and phospholipids, and is involved in energy transfer. A deficiency in phosphorus results in excessive abscission of old leaves, bronze, lusterless leaves and sparse bloom. Some typical fruit disorders have also been described (Chapman, 1968). Excess of P has been found to accentuate zinc deficiency.

Magnesium (Mg) is part of the chlorophyll molecule, and an activator of photosynthesis and respiration. Deficiency causes a characteristic chlorosis of the foliage (Table 5.5). It is a highly mobile element;

Table 5.5 *Functions and deficiency symptoms of macro and micronutrients*

Mineral nutrient	Major functions	Deficiency symptoms
N	Structural component of proteins nucleic acids and chlorophyll	Uniformly light green to yellowish green leaves; often reduced growth, decreased flowering and fruiting
P	Part of nucleoproteins and phospholipids; involved in high energy processes	Mostly not clearly manifested; leaf symptoms as in N deficiency; lusterless, bronze leaves; sometimes excessive abscission of leaves; more acid fruit
K	Involved in carbohydrate metabolism; acts as an osmoticum; involved in stomata closure	Mainly fruit symptoms; smaller fruit, peel thin, creasing. Sometimes smaller leaf, curling
Mg	Constituent of chlorophyll molecule; enzyme activator	Yellow blotches along midrib on mature leaves, till the pattern covers most of the leaf, apart from a characteristic delta shaped dark green area at the base
Zn	Involved in synthesis of the auxin IAA; chloroplast development	Chlorotic pattern on new growth; considerably smaller leaves, short internodes; in interveinal area characteristic light green to yellow mottle against dark green background
Mn	Involved in the splitting of water in light reactions; enzyme activator	Light green mottle in areas of leaf; band along midrib, major veins green; less sharply defined chlorosis than Fe deficiency
Fe	Involved in chlorophyll biosynthesis; enzyme activator	Chlorotic leaf pattern, first on young leaves; leaves light green, cream or yellow apart from fine network of midrib and veins, remaining green
Cu	Involved in oxidation–reduction systems; ascorbic acid oxidase, polyphenol oxidase	Enlarged dark leaves on willowy branches; brownish gum between bark and wood; dieback of twigs from tip; corky lesions on fruit peel
B	Involved in pollination and fertilization; carbohydrate metabolism	Small, hard, misshapen fruit; thick peel; pockets of brownish gum in peel and core
Mo	Part of NO_3 and NO_2 reducing system	Large interveinal yellow spots on older leaves (both surfaces); gum in chlorotic areas or underside of leaf

Table 5.6 *Mineral composition of tree parts of Marsh Seedless grapefruit*

	N		P		K		Ca		Mg	
	% in dry matter	% of total	% in dry matter	% of total	% in dry matter	% of total	% in dry matter	% of total	% in dry matter	% of total
Leaves	2.41	17.80	0.14	8.32	2.46	19.40	2.84	12.90	0.14	10.25
Immature fruits	1.56	5.72	0.18	5.00	1.80	7.05	0.21	0.41	0.11	4.17
Spring growth	1.18	1.32	0.17	1.67	1.63	1.89	1.79	1.23	0.11	2.08
Limbs up to ½ inch diameter	0.67	4.41	0.09	5.00	0.66	4.71	1.72	6.95	0.06	4.17
Limbs ½–2½ inches diameter	0.41	6.61	0.05	6.67	0.46	8.04	1.13	11.30	0.05	8.34
Old limbs	0.38	14.52	0.04	11.65	0.40	16.26	1.26	29.70	0.07	25.00
Trunk	0.47	6.61	0.05	5.00	0.38	5.68	1.35	11.58	0.08	10.25
Fibrous roots	1.02	20.24	0.21	31.70	1.13	23.60	0.43	5.21	0.10	18.80
Large roots	0.87	22.66	0.13	25.50	0.46	12.78	1.29	20.75	0.07	16.65

Modified from Chapman, H. D. (1968), after Barnette *et al.* (1931).

deficiencies start on basal leaves. They are especially noted in seeded citrus cultivars, because of translocation of Mg to the seeds. Magnesium is involved in the activation of several enzymes and maintenance of ribosomes.

Zinc (Zn) is a micronutrient. Next to nitrogen, zinc deficiency is perhaps the most widespread nutritional disorder in citrus, occurring under a wide range of soil conditions and environments. Deficiency symptoms are easily identified by characteristically mottled leaves, highly reduced leaf size, and often dieback of twigs and small misshapen fruit. Zinc is essential for the functioning of many enzymes, as well as the synthesis of tryptophan, a precursor of indoleacetic acid (IAA). Zinc deficiency causes a reduction in RNA synthesis and ribosome stability.

Foliar deficiency symptoms of both manganese (Mn) and iron (Fe) are chlorosis of leaves. Manganese deficiency symptoms occur both on young and old leaves. Iron deficiency starts on apical leaves. Chlorosis in Mn deficiency resembles Zn deficiency, but is usually less extreme. Severe Fe deficiency shows incomplete yellowing of leaves. Mn also activates several enzyme systems and is required in respiration and photosynthesis. Iron is involved in chlorophyll synthesis and is part of certain enzyme systems. It is involved in the reduction–oxidation process in photosynthesis and respiration.

Copper (Cu) deficiency symptoms are large, dark green leaves, gum pockets in woody tissue and between wood and bark, dieback of terminal shoots and multiple buds. Copper is part of the oxidation–reduction systems, such as ascorbic acid oxidase, polyphenol and laccase oxidases. It is part of plastocyanin (chloroplast enzyme).

While boron (B) deficiency has been a problem in some citrus orchards, excess B in the plant is even more common (see Table 5.7), especially in some irrigated regions and in soil areas high in boron. Boron deficiency is manifested by abnormal abortion of young fruits, albedo discoloration of fruit and dieback of growth. Symptoms are, however, not always highly specific. The concentration range of B in plants between deficiency, normal concentration and toxicity is rather narrow. Brown pustules on dark green leaves are symptoms of toxicity. Boron, not readily translocated, appears to be required for sugar translocation. It is involved in pollen tube elongation, and in cell division in root and shoot apices.

Molybdenum (Mo) has been established as an essential element by Arnon and Stout (1939). Yellow spot of citrus leaves in Florida was shown to be caused by molybdenum deficiency. Water-soaked areas appear on the leaves, becoming yellow. Eventually, gumming on the under surface appears, turning black in appearance. Mo is involved in the nitrite and nitrate reducing systems acting as an electron carrier.

Table 5.7 *Leaf analysis standards for citrus in Florida (4–6-month-old leaves on nonfruiting terminals)*

Element	Deficient	Low	Optimum	High	Excess
N (%)	<2.2	2.2–2.4	2.5–2.7	2.8–3.0	>3.0
P (%)	<0.09	0.09–0.11	0.12–0.16	0.17–0.29	>0.30
K (%)	<0.7	0.7–1.1	1.2–1.7	1.8–2.3	>2.4
Ca (%)	<1.5	1.5–2.9	3.0–4.9	5.0–6.9	>5.0
Mg (%)	<0.20	0.20–0.29	0.30–0.49	0.50–0.70	>0.80
Cl (%)	?	?	0.05–0.10	0.11–0.20	>0.20
Mn (ppm)	<17	18–24	25–100	101–300	>500
Zn (ppm)	<17	18–24	25–100	101–300	>300
Cu (ppm)	<3	3–4	5–16	17–20	>20
Fe (ppm)	<35	36–59	60–120	121–200	>200
B (ppm)	<20	21–35	36–100	100–200	>250
Mo (ppm)	<0.05	0.06–0.09	0.10–1.0	2.0–5.0	>5.0

From Koo *et al.* (1984). Recommended fertilizer and nutritional sprays for citrus, IFAS Bull 536 D. University of Florida Agricultural Experiment Station.

Assessment of nutrient deficiencies and toxicities

Soil tests and the determination and evaluation of the nutritional status of plant tissues provide important tools to assess deficiencies, substantiate diagnostics and help solve nutritional problems. Soil tests are of primary value for the determination of soil pH and the need to apply calcium to acid soils. They may be also of value in assessing P levels, in determining NO^-_3 levels as an indicator of N status and for the detection of toxic levels of Cu, B and salinity. Testing for other elements is being largely abandoned, partly due to the inherent sampling errors arising from having few samples represent a considerable volume of non-uniform soil. A more reliable means of assessing the nutrient status of citrus orchards is the use of plant tissue analysis. Analysis of leaf tissue provides the main method for determining the actual mineral content of the plant. Recently, some interest has also arisen in the mineral analysis of the fruit juice.

The leaf-analysis approach integrates the effects of many factors into one by determining the concentration of a given element in the leaf. When multiple deficiencies occur, the final diagnosis will depend on leaf analysis. It is also particularly useful in determining the tree's current nutritional status. However, leaf age, the presence of fruit, leaf position, and the tree variation and rootstock all affect the concentration of elements in leaves. Proper sampling is extremely important and is mostly done by analytical laboratory staff. In Florida, leaf samples consist of 100 or more spring flush leaves, collected in late summer, from nonfruiting twigs from at least 20 trees. Leaf analysis standards for citrus based on 4–6-month-old spring cycle leaves from nonfruiting terminals are shown in Table 5.7

As enzyme activity may bear a relation to the levels of mineral nutrients, the use of enzyme activity has been suggested as a measure of the adequacy (and deficiency) of the levels of certain elements. This approach is already in use for diagnosing Fe and Mn deficiency; peroxidase enzyme activity is lowered below normal by Fe deficiency, while it is increased with Mn deficiency (Bar-Akiva, 1961). A similar approach is being investigated with nitrate reductase (for N) (Bar-Akiva and Sternbaum, 1965), ascorbic acid oxidase (for Cu) and carbonic anhydrase (for Zn). Estimation of leaf nitrates, besides being simpler than N determination, has also proven highly indicative in some cases (Bar-Akiva, 1974).

A special statistical method called DRIS (diagnosis and recommendation integrated system) (Davee et al., 1986; Beverly et al., 1984) is also used for evaluating leaf analysis results in order to formulate fertilizer recommendations. It takes into account relations between different elements, allowing determination of the nutrient or nutrients that are the limiting factor.

Requirements for mineral nutrients

The concentration of most mineral nutrients in leaves is highest early in the season, decreasing during the season. Decreases in nitrogen are attributed to a dilution effect, because the total amount of N shows no decrease. Concentration of K in the leaves decreases noticeably during the season. Concentrations of Ca and Mg remain on the same level and often

Table 5.8 *Amounts of mineral elements removed by 40 tons of oranges*

Element	Amount removed by 40 tons of fresh fruit (kg)
N	47.2
P	10.5
K	102.2
Ca	41.8
Mg	7.6
Zn	0.026
Cu	0.016
Fe	0.112
B	0.104

Adapted from Chapman (1968).

increase during growth. Concentrations of phosphorus decrease some-what, and those of calcium, boron, iron and manganese increase, while zinc and copper change little with age. The influence of age of orange leaves on concentrations of nitrogen in leaves is shown in Figure 5.6.

Evidence indicates that the optimum nutritional level of a fertilizer element is rather similar for trees of a given variety, regardless of soil and climate. However, the ability of citrus trees to obtain and use elements in adequate quantities is influenced by numerous factors, including soil, climate, crop, rootstock and scion/rootstock compatibility.

A good crop of oranges (40 tons ha^{-1}) removes considerable quantities of some minerals from the soil (Table 5.8). Potassium is removed in the largest amount, followed by N and Ca. Since potassium accounts for over 40% of the total mineral content of citrus fruit, this may be one of the main reasons why shortage of potassium decreases fruit size. Potassium content tends to be lower during a heavy crop year. Investigations in South Africa (Du Plessis and Koen, 1988) have demonstrated that high N/K ratios decrease yield and fruit size in Valencia oranges.

The seasonal demand for mineral nutrients is greatest during the period of flowering, fruit set and maximum shoot growth.

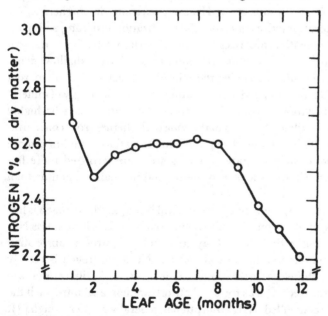

Figure 5.6 Influence of age of orange leaves on concentrations of nitrogen in leaves from nonfruiting terminals. From Reuther, Walter, ed. (1973). *The Citrus Industry*, Vol 3. Berkeley, University of California, Division of Agricultural Sciences, Publication 4014. Reproduced with permission

Fertilization of citrus orchards

Many approaches have been suggested for formulating fertilization programs for citrus groves; replacement of nutrients removed by fruit,
symptomatology, greenhouse and field experiments, and soil and leaf
analysis. Deficiency characteristics of Zn, Mn, Mg, Fe, Mo, Cu are well
characterized, but prevention of deficiencies is generally preferable to
treatment. Interpretation of field experiments using leaf analysis from the
citrus orchard seems to be applicable for a rather wide range of environments.

Soil conditions affect fertilizer needs. Sandy soils with a low capacity to
bind nutrients are more subject to leaching of minerals below the root zone
than heavy soils. Soil pH largely determines the availability of many
mineral ions. In acid soils (pH<6.0) excess hydrogen ions may replace
cations (Ca^{++}, K^+, Mg^{++}, Na^+). In highly acid soils solubilities of Mn^{++}
and Al^{+++} are high, often with toxic effects. Application of lime ($CaCO_3$) is
usually part of a program to increase the pH of the soil. On highly alkaline
soils (pH>8.0 or 8.5) mineral elements, especially micronutrients, may
become unavailable to the plant. The ideal pH would be between 6.0 to 7.0,
with still satisfactory results up to pH 8.0 or even somewhat higher.

Of the cultural practices affecting the fertilization program, soil management and irrigation practices play an important role. Orchards irrigated with a drip or low-volume irrigation system have a shallow, dense
root system concentrated under the emitters. Trees with limited root
systems are also more dependent on a rather continuous supply of mineral
nutrients. Fertigation (fertilization with drip irrigation) can be highly
efficient as the fertilizer is placed in the zone of the highest root concentration. This results in a reduced requirement for fertilizers and in a more
continuous period of fertilization. Young trees are fertilized more frequently (four to six times a year, or more) and in smaller amounts than
mature trees.

Sooner or later, supplemental nitrogen will be required. Nitrogen is lost
by leaching, volatilization, runoff and crop removal. In Florida, less N is
absorbed in winter than from May to October. Nitrogen absorption
depends on the temperature of soil and air. Nitrate nitrogen is readily
absorbed and, therefore, most commonly it is nitrate fertilizers that are
applied to citrus trees. Citrus trees will absorb ammoniacal nitrogen if the
pH is properly controlled. NH_4 absorption is generally greater at high pH.
Under optimum conditions, a major portion may be nitrified to nitrate in
two to three weeks. Soil conditions are not always optimal for the
nitrification process. A shortage of nitrogen in citrus trees is likely to be
most critical prior to and during blossoming, and during 'June drop'

(Chapman and Parker, 1942). Foliar applications of urea or KNO_3 are also used, especially after cold, wet winters.

The timing and number of N applications will depend on tree and soil conditions. However, in the northern hemisphere, fall applications, preferably after fruit color break, are given to encourage the build-up of reserves in the tree and to increase the availability of N for sprouting and flowering. Some authorities do not recommend applications after July (in the northern hemisphere) because of a possible delay in fruit color break. Increasing N levels, while increasing fruit size (to a lesser extent than does K), may also decrease juice content and ascorbic acid level. As nitrate is highly water soluble, excessive nitrogen applications may contribute to groundwater contamination.

Most soils have a comparatively high content of total K; only a small percentage is in a readily available form (such as exchangeable K absorbed on colloids or ions in solution). Fertilization with K can be effective (though not acting immediately) at any time during the growing season. Potassium fertilization practice will depend also on its effect on fruit maturity, color, characteristics and size. Increased K levels enlarge fruit size, reduce fruit creasing and may increase fruit coarseness and thickness.

Phosphorus deficiency occurs quite rarely, though it is quite prevalent in South Africa and has been found on acid soils in California. Symptoms are rarely striking; leaf and soil analyses are much more indicative. Vesicular arbuscular mycorrhizae (VAM) associated with the fungus species *Glomus* have been found to increase P uptake by certain rootstocks, especially in citrus nurseries (Nemec, 1978).

Application of most micronutrients and of Mg in alkaline soils is usually done by nutritional sprays (Swietlik and Faust, 1984). Zn sprays in the foliage are regularly applied in many orchards. B is applied to the soil or as a micronutrient spray. Mn and Mo sprays are usually applied after observation of a deficiency. Iron is often unavailable to the trees in high pH soil, under waterlogged conditions, in soils very low in organic matter and in highly calcareous soils. Organic combinations of iron (chelated iron) are usually applied to the soil. While iron ethylenediamine tetra-acetic acid (FeEDTA) is satisfactory on soils with acid pH, hydroxy-chelates such as iron ethylenediamine di-o-hydroxyphenyl acetic acid (FeEDDHA) are used on soils with an alkaline pH.

Use of plant growth regulators

Plant growth regulators play an important role in the control of developmental processes in agricultural plants. Since the early 1950s growth regulators have assumed an increasingly significant role in the management of orchards and in the postharvest handling of citrus fruit. Work in this area has appeared in hundreds of reports and has been subsequently summarized in useful review articles which provide an up-to-date view of this diversified area (Coggins and Hield, 1968; Monselise, 1979; Coggins, 1982; Hirose, 1982; Wilson, 1983; Davies, 1986; Harty and van Staden, 1988; Augusti and Almela, 1991).

The rationale underlying most uses of plant growth regulators is the intervention in plant processes via the endogenous hormonal systems that regulate plant development. Thus, most growth regulators are either analogues of native plant growth substances or compounds that inhibit or enhance the biosynthesis of native plant growth substances. The terms 'bioregulators' and 'plant biochemical regulators' (Gausman, 1991) are somewhat broader, including also chemicals which interfere with non-hormonal metabolic plant systems (e.g. tertiary amine compounds that affect carotenoid biosynthesis) (Gausman, 1991). The use of lead arsenate to increase fruit sweetness (Wilson, 1983) may come under this heading. Most of the abscission chemicals used to facilitate fruit harvesting function by causing superficial peel burn, followed by the production of wound ethylene (Wilson, 1983). These should also be classified as bioregulators. While certain growth regulators have been mentioned in Chapter 4, Table 5.9 summarizes most of the treatments commonly used in citriculture, relating them to the native hormonal systems that are presumably involved. Reference is made also to alternative agrotechnical treatments. It can be seen from Table 5.9 that the compounds used are mostly auxins, gibberellins and ethylene (or their effectors). Cytokinins and abscisic acid (ABA) have not yet found a use in citriculture.

As with other agrochemicals, the use of plant growth regulators may involve certain health hazards. In order to minimize the risk of growth regulator applications, efforts have been oriented recently toward improvement of the efficiency of treatments. Adjustment of the pH of treatment solutions, selection of appropriate surfactants and application under high relative humidity conditions improve uptake and enable considerable reduction of the amount applied (Goldschmidt and Greenberg, 1989; Greenberg and Goldschmidt, 1990).

Frost damage and prevention

Citrus, classified as a cold-tender evergreen, is highly vulnerable to freeze injury. Its ability to survive freezing temperatures ranks far below that of

Table 5.9 *Effects and major uses of plant growth regulators in citriculture*

Process	Principal endogenous system involved	Growth regulator treatments	Alternative agrotechnical treatments
Vegetative shoot growth	Gibberellins, auxins	Paclobutrazol and other growth retardants (−)	Root restriction, dwarfing stock/scion combinations
Flowering	Gibberellins	Paclobutrazol (+) Gibberellin A_3 (−)	Drought stress, girdling
Fruit set	Gibberellins	Gibberellin A_3 (+)[2]	Girdling
Fruit thinning	Ethylene/auxin	Synthetic auxins,[1,2] ethephon (+)	Manual thinning
Fruit growth	Gibberellins, auxin	Synthetic auxins (+)[2]	Fruit thinning, girdling, potassium sprays
Creasing of fruit peel	Gibberellins	Gibberellin A_3 (−)[2]	Potassium sprays
Fruit splitting	?	Synthetic auxins (−)	Potassium sprays, reduced irrigation during pulp growth
Preharvest fruit abscission	Ethylene/auxin	Synthetic auxins (−)[2]	—
Fruit peel coloration	Ethylene/gibberellins, cytokinins	Ethylene (+)[2,3] Gibberellin A_3 (−)[2]	—
Fruit button (calyx) abscission	Ethylene/auxin	Synthetic auxins (−)[2,3]	Heavy nitrogen fertilization
Fruit peel viability	Gibberellins, cytokinins/ethylene	Gibberellin A_3 (+)[2,3]	—

[1] Acting as inducers of ethylene production by plant tissues (Hirose, 1982).
[2] Widely recommended treatment.
[3] Used mainly postharvest.
(+) = promotion; (−) = inhibition.

northern woody species. Many of the world's major citrus-producing areas are prone to periodic cold-weather damage (as exemplified by Florida, where temperatures of $-5\,^{\circ}\mathrm{C}$ and below are experienced about once every seven years). In southern Japan, freeze hazards occur quite often during winter, though there is a considerable moderating influence of the Pacific Ocean. Severe freezes occur also in many subtropical, arid to semiarid regions. The terms 'radiative' and 'advective' are often used to define freezing conditions. A feature of radiative freeze is temperature inversion, denoting a layer of atmosphere in which temperature increases with height. Some use 'frost' for hoarfrost or for local occurrences and 'freeze' for a more general event resulting in advection of large masses of air at subfreezing temperatures.

Freezing temperatures cause rupture of cells, disruption of membrane function and irreversible denaturation of enzymes through dehydration. Damage occurs as water present in plant tissues freezes and expands in volume.

Lethal injury usually occurs between $-2.2\,^{\circ}\mathrm{C}$ and $-6.7\,^{\circ}\mathrm{C}$ (Young, 1969). Non-growing twigs and limbs of citrus may be injured at temperatures ranging from $-2\,^{\circ}\mathrm{C}$ to $-9\,^{\circ}\mathrm{C}$. It has been suggested that once ice has formed in citrus tissues, a critical time lapse, inversely related to minimum temperature, is required before irreversible injury occurs (Yelenosky, 1985). Succulent flowers and new growth (over 85% tissue water content) are most vulnerable to ice damage. Freeze-damaged fruit poses a considerable problem in maintaining fruit quality. While severely damaged fruit may drop, moderately damaged fruit may remain on the tree for several weeks. This involves drying out of juice sacs or part of the fruit. Lower juice content, lower TSS and also lower TA (total acidity) have been found in freeze damaged fruit. Fruit is usually damaged after four hours below $-2\,^{\circ}\mathrm{C}$ (Yelenosky, 1985), or after several hours at $-3\,^{\circ}\mathrm{C}$ (Mukai and Kadoya, 1994).

Citrus remains uninjured if ice does not form in the tissue. The fruit peel of many varieties is, however, chill sensitive and has been injured at temperatures above $0\,^{\circ}\mathrm{C}$ (Purvis, 1980). Citrus leaves are considered chill resistant. Chloroplast ATPase and ribulose bisphosphate carboxylase are reversibly inhibited at temperatures above freezing.

There are significant differences in cold-hardiness rating of citrus types and cultivars (Young and Hearn, 1972). *Poncirus trifoliata* ranks highest in cold hardiness, followed by *Fortunella*. Mandarin cultivars such as Changsha and Satsuma rank highest in cultivated citrus, followed by sour orange, mandarins, sweet orange, grapefruit, lemon, lime and citron. Citron, lime and lemon tend to grow continuously and to flower and set fruit in cool weather, which increases their susceptibility to frost. Actively growing trees are less cold hardy than non-growing ones, with cold hardiness usually greater in midwinter than in fall or spring. Citrus trees generally stop growing at air temperatures below 12 °C. Cessation of growth during late fall and winter aids in predisposing citrus trees to becoming cold-hardened. NMR analysis showed cold-hardened Satsuma leaves to have a higher percentage of unfrozen water than nonhardened leaves during freezes (Yelenosky, 1985). *Poncirus* rootstock contributes to hardiness in the scion (Satsuma on *Poncirus trifoliata* has very high cold tolerance). For breeding for improved cold hardiness see Chapter 6.

Natural mechanisms protecting against freeze damage have evolved in citrus, contributing to survival during severe freezes. Adaptive physiological and metabolic changes are involved, resulting in different cellular composition and physical relationships.

Citrus cold hardening is highly influenced by temperature. Light is also an important factor, as well as water relations and anatomical differences. Prehardening low temperatures cause an increase in solute accumulation and decreased water uptake. Bud dormancy inducing temperatures have been studied by Young (1969). Though all of the major categories of plant components have been implicated in cold hardness in citrus, lipid metabolism is particularly involved in adaptation to the environment and in the function of plant membranes. Increases in linoleic acid as well as in the triglyceride level have been noted in cold-hardened citrus (Nordby and Yelenosky, 1982).

Yelenosky (1985) envisages a three-stage process.

1 Predisposition by good tree health.
2 Prehardening induced by cool, nonfreezing temperatures.
3 Actual cold hardening by cellular responses to cooler temperatures, depending on genetic traits.

Water in plant tissues remains in a liquid state at temperatures sometimes well below 0 °C in the absence of a catalyst for ice formation. The phenomenon has also been referred to as supercooling, constituting an effective freeze avoidance adaptation in plant tissues (Levitt, 1980a). The inherent ability to supercool is a main contributor to the avoidance of major freezing effects in many citrus areas. Most citrus species rarely supercool below −10 °C, while *Poncirus* may supercool at appreciably lower temperatures (Yelenosky, 1985). Anderson *et al.* (1983) found that Satsuma leaves tolerate −10 °C, foliage of lemon and lime was killed at −4 °C, with sweet orange and grapefruit intermediate. Differences in hardiness were explained by the amount of frozen water tolerated at the killing temperature. Reduced ice formation could not be attributed to osmotic effects.

Ice nucleation active bacteria (INA), mostly *Pseudomonas* and *Erwinia*, have been found to be implicated in inducing ice formation in plants during freeze conditions (Lindow, 1982). The importance of INA bacteria has not been fully established in the major freeze-prone citrus areas. Isolations from citrus plants in Israel resulted in two bacteria in active ice nucleation at −2.5 °C (Yankofsky *et al.*, 1981). *P. viridiflava* was found most often on citrus in Florida. Ice nucleation has been attributed to a nondiffusable proteinaceous substance in the outer cell membrane of the bacteria (Orser *et al.*, 1985). Efforts to reduce the number of INA bacteria on the plant included application of antibiotics and heavy metals prior to a freeze. Adoption of such practices is still very much limited because of lack of effective, reliable materials.

Passive frost protection involves the selection of site, scion and root-stock, and factors affecting cold hardiness, among others. Active freeze protection has been classified by Rieger (1989) into:

1   Addition of heat.
2   Mixing air above orchard with cooler air within (during inversion).
3   Conservation of heat.

Some consider sprinkling of water a further category.

Protective measures in citrus have been discussed by Turrell (1973). The sporadic nature of frost occurrences complicates adoption of high-cost frost-protection measures for mature trees. With cheap fuel oils, orchard heaters have been used to a large extent, especially in California. Nowadays, they are employed mainly in high-value citrus nurseries or where high-value fresh fruit is raised. Wind machines can be of a certain value in sites experiencing temperature inversion, but are rather in-efficient in highly windy situations. Irrigation is the most popular method of freeze protection in recent use in Florida (Smajtrla, 1993).

Operating microsprinklers continuously under tree canopies raises the temperature under the tree due to the ambient water temperature and by releasing heat of fusion as water freezes. Climatic factors, tree variety and degree of acclimation will affect the degree of protection achieved (Smajtrla, 1993).

Freezing water maintains the trunk and bud union near 0 °C, providing protection from subfreezing temperatures. With water constantly applied to the young canopy and ice continually being formed, protection has been maintained, even at temperatures of −8°C to −10°C. Operating microsprinklers continuously under mature tree canopies during freezes provided some protection (Parsons *et al.*, 1982), but water proved much less effective during advective freezes. The base of the tree can be protected but the top may remain largely unprotected.

Efficiency of microsprinklers for protection of young trees, especially when combined with tree wraps, proved very high (Davis *et al.*, 1987). Tree covers, supplemented by microsprinklers, raised canopy temperature by 7.8 °C (Jackson *et al.*, 1986). Winter-planted trees in Florida are protected by banking soil (low thermal diffusivity) or, more conveniently, by using insulating wraps, preferably up to the scaffold (Jackson, 1991).

Biotic stress: pests

Citrus pests and their management

Wherever citrus is grown, it is attacked by numerous pests – mostly insects and mites, but also snails and various vertebrates. A list of the more important arthropod groups is presented in Table 5.10. For more extensive lists of citrus pests and discussions of their life histories see Quayle (1938a), Bodenheimer (1951), Ebeling (1959), Chapot and Delucchi (1964), Talhouk (1975), Bedford (1978), Knapp (1987), Reuther *et al.* (1989), *Proceedings of the International Citrus Congresses* and textbooks of agricultural entomology.

In its native range in the Far East, where citrus is grown quite extensively for local consumption, various lepidopterous and coleopterous species are usually the most injurious, whereas homopterous pests are generally kept under effective control by natural enemies. On the other hand, in large-scale, export- and industry-oriented citriculture Homoptera are usually major pests, along with mites, fruit flies and thrips.

The factors responsible for pest problems in modern citriculture are many and varied, but they can be grouped into three main categories: invasions, ecological changes and socio-economic changes (Rosen, 1986).

Invasions are, unfortunately, all too common. Intercontinental travel and commerce, and in particular the deliberate transfer of citrus plants, have provided ample opportunity for pest species to migrate and become established in new habitats. A majority of the serious pests in any modern citrus ecosystem is species of foreign origin, and invasions repeatedly occur even nowadays. Thus, the Mediterranean fruit fly, *Ceratitis capitata*, has invaded both California and Florida several times during the present century, whereas recent invasions into Israel have resulted in the establishment of such notorious citrus pests as the spirea aphid, *Aphis spiraecola*, the citrus whitefly, *Dialeurodes citri*, the bayberry whitefly, *Parabemisia myricae*, and the citrus red mite, *Panonychus citri*. Regulatory control in the form of strict quarantines is of primary importance in preventing pest problems.

Table 5.10 *Major groups of arthropod citrus pests*

Class/Order/Family	Representative species/Common name	Nature of injury
Acari		
Prostigmata		
Eriophyidae	*Eriophyes sheldoni* Ewing Citrus bud mite	Mites feed in buds, stunting trees and causing grotesque deformations of fruit and leaves
	Phyllocoptruta oleivora (Ashmead) Citrus rust mite	Mites feed on epidermal cells, causing silvering of yellow fruit and russetting of oranges
Tarsonemidae	*Polyphagotarsonemus latus* (Banks) Broad mite	Silvering of lemons, leaf curl
Tenuipalpidae	*Brevipalpus phoenicis* (Geijskes) Citrus flat mite	Chlorosis of fruit and leaves, galls on stems of seedlings
Tetranychidae	*Panonychus citri* (McGregor) Citrus red mite	Defoliation, fruit drop, discoloration, reduction of fruit size
Insecta		
Homoptera		
Aleyrodidae	*Aleurocanthus woglumi* Ashby Citrus blackfly	Whiteflies and blackflies suck the sap from leaves and shoots, draining the plant and producing large amounts of honeydew, on which sooty mold fungi develop, blocking photosynthesis and contaminating fruit. Defoliation and crop loss may follow
	Aleurothrixus floccosus (Maskell) Woolly whitefly	
	Dialeurodes citri Ashmead Citrus whitefly	
	Parabemisia myricae (Kuwana) Bayberry whitefly	
Aphididae	*Aphis spiraecola* Patch Spirea aphid	Aphids suck sap from young growth, causing leaf curl, blossom drop and sooty mold. Some species are vectors of tristeza and other diseases
	Toxoptera aurantii (Boyer de Fonscolombe) Black citrus aphid	

Coccidae	*Ceroplastes destructor* Newstead — White wax scale *Ceroplastes floridensis* Comstock — Florida wax scale *Ceroplastes rubens* Maskell — Pink wax scale *Coccus hesperidum* L. — Brown soft scale *Coccus pseudomagnoliarum* (Kuwana) — Citricola scale *Saissetia oleae* (Olivier) — Mediterranean black scale	Soft and wax scales suck sap from leaves and branches, producing copious amounts of honeydew and sooty mold. Defoliation, fruit drop and deadwood may result, and blemished fruit may be culled
Diaspididae	*Aonidiella aurantii* (Maskell) — California red scale *Chrysomphalus aonidum* (L.) — Florida red scale *Chrysomphalus dictyospermi* (Morgan) — Dictyospermum scale *Lepidosaphes beckii* (Newman) — Purple scale *Parlatoria pergandii* Comstock — Chaff scale *Selenaspidus articulatus* (Morgan) — Rufous scale *Unaspis citri* (Comstock) — Citrus snow scale *Unaspis yanonensis* (Kuwana) — Arrowhead scale	Armored scale insects suck the contents of cells on all parts of the tree, causing defoliation, fruit drop, dying back of twigs and branches, distortion and discoloration of fruit, and culling of infested fruit. Severe attack may result in death of trees

Table 5.10 continued

Class/Order/Family	Representative species/Common name	Nature of injury
Insecta		
Homoptera		
Margarodidae	*Icerya purchasi* Maskell Cottony cushion scale	Sucking of sap from leaves and twigs may cause defoliation, fruit drop and heavy sooty mold
Pseudococcidae	*Planococcus citri* (Risso) Citrus mealybug *Pseudococcus calceolariae* (Maskell) Citrophilus mealybug *Pseudococcus citriculus* Green Citriculus mealybug	Mealybugs suck sap, producing honeydew and sooty mold, sometimes attracting lepidopterous fruit pests. Feeding at fruit stems may cause fruit drop
Psyllidae	*Trioza erytreae* (Del Guercio) Citrus psylla	The principal vector of greening disease; nymphs cause leaf galls
Cicadellidae	*Empoasca citrusa* Theron Green citrus leafhopper	Leafhoppers suck fruit, causing dark or chlorotic spotting, even fruit drop
Heteroptera		
Coreidae	*Leptoglossus phyllopus* (L.) Leaf-footed bug	Sucking of sap causes wilting of growth tips, and sometimes fruit loss
Pentatomidae	*Rhynchocoris humeralis* (Thunberg) Citrus green stinkbug	Stinkbugs suck sap, causing serious crop loss
Thysanoptera		
Thripidae	*Frankliniella occidentalis* (Pergande) Western flower thrips *Scirtothrips citri* (Moulton) Citrus thrips	Thrips feed on epidermal cells, scarring fruit, destroying buds, dwarfing and distorting leaves
Orthoptera		
Acrididae	*Schistocerca americana* (Drury) American grasshopper	Nymphs feed on citrus leaves and on the peel of orange fruit, causing defoliation of young trees and blemished fruit

Coleoptera		
Bostrychidae	*Apate monachus* F. Black giant bostrychid	Adult beetles bore tunnels in trunk and branches
Buprestidae	*Agrilus occipitalis* Eschscholz Citrus bark borer	Larvae mine under bark of branches and roots, killing young trees
Cerambycidae	*Melanauster chinensis* Förster Citrus trunk borer	Larvae bore into trunk, branches and roots, killing trees. Adult beetles feed on foliage
Chrysomelidae	*Throscoryssa citri* Maulik Black-and-red leaf miner	Larvae skeletonize foliage
Curculionidae	*Pachnaeus litus* (Germar) Citrus root weevil *Sciobius granosus* Fahraeus Citrus snout beetle	Larvae feed on roots, causing decline of trees and reduced yields. Adult beetles feed on foliage
Diptera		
Tephritidae	*Anastrepha ludens* (Loew) Mexican fruit fly *Bactrocera dorsalis* (Hendel) Oriental fruit fly *Bactrocera tyroni* (Froggatt) Queensland fruit fly *Ceratitis capitata* (Wiedemann) Mediterranean fruit fly	Fruit flies oviposit in ripening fruit, the punctures and developing larvae causing decay and fruit drop. Serious crop loss may result. Many species are on quarantine lists, preventing export of citrus and other fruit
Hymenoptera		
Formicidae	*Atta sexdens* (L.) Leaf-cutting ant *Iridomyrmex humilis* (Mayr) Argentine ant	Many ants feed on honeydew, fostering homopterous pests and decimating their natural enemies. Leaf-cutting ants feed on fungi grown on cut leaves, and may cause serious damage to foliage

Table 5.10 *continued*

Class/Order/Family	Representative species/Common name	Nature of injury
Insecta		
Lepidoptera		
Geometridae	*Ascotis selenaria reciprocaria* Walker Citrus looper	Caterpillars feed on fruit, leaves and blossoms
Lyonetiidae	*Phyllocnistis citrella* Stainton Citrus leaf miner	Caterpillars mine leaves, may kill young trees
Metarbelidae	*Indarbela quadrinotata* (Walker) Bark-eating borer	Caterpillars bore into bark
Noctuidae	*Othreis cajeta* (Cramer) Fruit-piercing moth	Adult moths pierce fruit, causing fungal decay and fruit drop
Olethreutidae	*Cryptophlebia leucotreta* (Meyrick) False codling moth	Caterpillars bore into ripening fruit, causing decay and fruit drop
Papilionidae	*Papilio demolens* L. Lemon butterfly	Caterpillars feed on leaves, may defoliate young trees
Pyralidae	*Citripestis sagittiferella* Moore Citrus moth borer	Caterpillars burrow in rind and may penetrate fruit, causing blemishes and fruit drop. Some are secondary pests, attracted to mealybug colonies
	Cryptoblabes gnidiella (Millière) Honeydew moth	
	Ectomyelois ceratoniae (Zeller) Carob moth	
Tortricidae	*Argyrotaenia citrana* (Fernald) Orange tortrix	Caterpillars feed on young foliage and fruit, and may also bore into ripening fruit
Yponomeutidae	*Prays citri* Millière Citrus flower moth	Caterpillars feed on flowers, reducing fruit set
	Prays endocarpa Meyrick Citrus rind borer	Caterpillars burrow in fruit, causing unsightly swellings and scars

Ecological changes are another major cause of pest outbreaks. The history of agriculture has been the history of constant ecological change. By creating monocultures and high-yielding plant varieties, by eliminating competitors or natural enemies through various agrotechnical practices, humans have often inadvertently created conditions favorable to certain species and have induced a manyfold increase in their populations. Recent changes in the citrus fauna of Israel may serve to illustrate this point. The California red scale, *Aonidiella aurantii*, had been known in the past mainly as a pest of young citrus groves in Israel, having usually been competitively displaced from mature groves by the Florida red scale, *Chrysomphalus aonidum*. When the Florida red scale was brought under complete biological control by an introduced parasitoid, the California red scale – unfortunately, much more difficult to control – became a major pest of both young and mature groves. Then, as large citrus acreages planted in the 1950s came to maturity, the chaff scale, *Parlatoria pergandii*, a habitual pest of older groves, gradually became more injurious.

Socio-economic changes may be as important in affecting pest status as are actual changes in the physical environment of the citrus ecosystem. Economic thresholds for pest management are determined by the market value of the crop as compared with the cost of available control measures, as well as by consumer habits and taste. Changes in public attitude towards pest-blemished citrus fruit, for instance, may drastically lower the thresholds and thus cause a hitherto insignificant organism to be considered an economic pest, even though its actual population density may have remained unaltered.

Pest management techniques

Chemical control of pests has been a salient feature of citriculture ever since the early days of HCN fumigation, botanical insecticides and mineral oil sprays, but the modern era of chemical control began some 50 years ago, with the advent of DDT and numerous other synthetic organic pesticides. By and large, these modern pesticides have provided citrus growers with effective, and usually quite reliable, means for controlling arthropod pests. However, this has been a mixed blessing. Not only do chemical pesticides, at best, provide only temporary relief from pest problems, but their massive overuse and frequent misuse have often resulted in grave problems, including those of ever-increasing cost, resistance and toxicity.

The high price of modern pesticides, and the large volumes required for adequate coverage of a mature citrus grove, have threatened the profitability of citriculture in many areas. This has been complicated by loss of pesticides due to rapid development of resistance in pest populations, known in citriculture almost since the turn of the century, when citrus scale insects were found to be resistant to HCN (Quayle, 1938b). Many of the modern pesticides are highly toxic to humans, and numerous poisoning accidents have occurred among pest-control operators and others exposed to them. Continued use of such pesticides on citrus may cause serious marketing problems, as several importing countries have recently imposed strict residue tolerances. Phytotoxicity to citrus itself is also quite common, certain varieties such as the Shamouti (Jaffa) orange being exceedingly sensitive to oil/organophosphorus combinations and various other formulations. Finally, broad-spectrum pesticides applied to citrus or to adjacent crops have been notorious for disrupting the biological equilibrium in many citrus ecosystems, causing severe resurgences of 'old' pests and upsets of 'new' pests through decimation of their natural enemies (DeBach and Bartlett, 1951; DeBach and Rosen, 1991).

In view of these drawbacks, it is evident that if chemical pest control is to play a significant role in the future of citriculture, it will have to undergo some drastic changes. It needs to become more selective, much safer environmentally, and also considerably less expensive. Some conventional pesticides are physiologically selective, being much less toxic to non-target organisms than to certain pests. Tests of toxicity to beneficials have led to the replacement of sulfur, once the prevalent means for controlling mites on citrus, with more selective acaricides (Rosen, 1967a,b).

These changes are not always easy. Mineral oils, for instance, are highly selective and are therefore recommended for the control of citrus scale insects (Riehl, 1990). However, their effective use requires complete coverage of the trees and careful timing of applications, and may be problematic when several species infest the same grove. Also, mineral oils may retard the development of yellow fruit color in early-ripening citrus varieties, and may cause severe injury, including defoliation and fruit drop, to some sensitive varieties during the summer. Citrus growers often turn to supposedly more effective, but certainly less selective, organophosphorus and carbamate pesticides.

Some of the new insect growth regulators appear to be highly effective against scale insects but rather harmless to their parasitoids (Darvas and Varjas, 1990). However, even a broad-spectrum pesticide may be applied in ways that would render it ecologically selective. The use of the organophosphorus pesticide malathion in bait sprays against the Mediterranean fruit fly may serve as a good example. Aerial strip sprays of very small amounts – one liter per hectare – of poison bait, containing protein hydrolysates as a powerful attractant for female flies and malathion as a poison, have proven to be far superior in effectiveness, and much less disruptive to the citrus ecosystem, than full-coverage applications of chlorinated hydrocarbons (Roessler, 1989).

In spite of the undisputed benefits of chemical control, attention has been increasingly focusing on the development of alternative methods. An impressive array of selective cultural, mechanical, physical, autocidal and biological techniques is available.

Cultural controls may include the use of pest-resistant plant cultivars, as well as various agrotechnical practices such as clean cultivation, tillage, water and fertilizer management, sanitation and timing of harvest.

Resistant plant cultivars are perhaps the ideal means of controlling a pest. Provided, of course, that such cultivars are not inferior in any other way, their introduction is by far the safest, most effective and most economical control method, and is highly compatible with all other methods. Regrettably, although there seems to be ample evidence that certain citrus species or varieties are resistant to such serious pests as the California red scale, *Aonidiella aurantii* (Compere, 1961) or the citrus leaf miner, *Phyllocnistis citrella* (Singh and Rao, 1980), very little progress has actually been made in citriculture in this promising direction. More recently, Greany (1989) suggested using gibberellic acid to increase the resistance of citrus fruit to fruit flies.

Some agrotechnical techniques of pest control have been employed in citriculture for centuries. Early harvesting of citrus fruit, for instance, may help citrus growers evade the spring peak of Mediterranean fruit fly infestation (Bodenheimer, 1951), whereas sanitation measures – eliminating dying branches and trees that serve as breeding sites – may prevent infestations of the black giant bostrychid, *Apate monachus* (Avidov and Harpaz, 1969). For an extensive review of other cultural control methods used in citriculture see Nucifora (1986).

Usually highly selective, cultural controls are often effective, safe, and compatible with all other practices. Their simplicity and low cost make them especially suitable for developing countries. However, for effective implementation, these methods require thorough knowledge of the biology and ecology of the target pests, indeed of the entire agro-ecosystem, and may differ from one area to another. Thus, clean cultivation of citrus groves is recommended in Florida as a preventive measure against the American grasshopper, *Schistocerca americana* (Knapp *et al.*, 1987), whereas planting a ground cover or 'green manure' crop in the groves is recommended in China as a means of augmenting predators of the citrus red mite, *Panonychus citri* (DeBach and Rosen, 1991). Indeed, much more research emphasis should be placed on cultural control methods in citriculture.

Mechanical and physical controls may range from extraction of wood borers with a hooked wire, or bagging fruit to prevent access of fruit flies, to highly sophisticated – but usually still experimental – uses of various forms of electromagnetic energy. Sometimes a simple method may prove highly effective. A fine example in this category is the use of high-pressure rinsing to remove scale insects and sooty mold from citrus fruit in the packinghouse (Bedford, 1990). This simple technique may eventually revolutionize pest-control practices in citriculture, permitting growers to raise economic thresholds in the grove to much higher levels. Another technique is the use of refrigeration to prevent the passage of the Mediterranean fruit fly with exported citrus fruit. By shipping the fruit under controlled temperatures in refrigerated boats, every single immature fly is exterminated during the voyage.

Autocidal controls include the sterile-insect technique and other genetic control methods, as well as the use of pheromones in pest management.

The sterile-insect technique, based on mass-releasing sterilized, laboratory reared insects (usually males) in order to reduce the chances of wild females mating with fertile males, has led to some spectacular projects in citriculture. Most notably, 'Programa Moscamed', releasing more than 500 million irradiation-sterilized flies per week, resulted in the Mediterranean fruit fly having been declared eradicated from Mexico by 1982, and is now continuing in Guatemala (Linares and Valenzuela, 1993). However, the applicability of this sophisticated technique is rather limited to fruit flies and a few other insect groups. Ecologically safe but rather expensive, it should be tried whenever feasible but should not be promoted out of proportion to its actual potential in citrus pest management (Rosen, 1986; DeBach and Rosen, 1991).

Pheromones, on the other hand, have come into extensive use not only in monitoring pest populations but also in actual control, and have already accounted for significant reductions in the use of broad-spectrum pesticides in a variety of agro-ecosystems. The development of an effective control program for the citrus flower moth, *Prays citri*, on lemons in Israel is an example. From an earlier method of population monitoring by traps baited with virgin females, a sophisticated system has been perfected whereby sticky traps baited with a synthetic female sex pheromone are used for mass trapping of male moths. This 'male annihilation' results in the females remaining unfertilized, and effective control is achieved without resort to chemical pesticides (Sternlicht *et al.*, 1978).

Other attractants have also figured prominently in pest management programs on citrus, especially against fruit flies. Trimedlure, a potent synthetic attractant for male flies, has been in extensive use in monitoring

Mediterranean fruit fly populations (Leonhardt *et al.*, 1984), whereas baits containing protein hydrolysates have been employed in actual control of this major pest of citrus (Roessler, 1989), as mentioned earlier. Male annihilation with another lure, methyleugenol, was used successfully to eradicate the Oriental fruit fly, *Bactrocera dorsalis*, from the Ryukyu Islands (Cunningham, 1989).

Biological control, i.e. the utilization of natural enemies – parasitoids, predators and pathogens – has been more successful on citrus than in any other major cropping system. Although the high proportion of successes on citrus may simply reflect the fact that more biological control efforts have been made on this than on any other crop, it is also partly due to the fact that most of the serious pests of commercial citrus are introduced species, notably Homoptera which, because of their sedentary nature and colonial habits, are perhaps more amenable to biological control than other groups of organisms (DeBach *et al.*, 1971). Landmark projects on citrus have included the control of the cottony cushion scale, *Icerya purchasi*, the California red scale, *Aonidiella aurantii*, the Florida red scale, *Chrysomphalus aonidum*, the purple scale, *Lepidosaphes beckii*, the rufous scale, *Selenaspidus articulatus*, the snow scale, *Unaspis citri*, the arrowhead scale, *Unaspis yanonensis*, the Mediterranean black scale, *Saissetia oleae*, the citrophilus mealybug, *Pseudococcus calceolariae*, the citriculus mealybug, *Pseudococcus citriculus*, the citrus blackfly, *Aleurocanthus woglumi*, the woolly whitefly, *Aleurothrixus floccosus*, the citrus whitefly, *Dialeurodes citri*, and the bayberry whitefly, *Parabemisia myricae*. Some of these outstanding projects have not only solved serious pest problems in citriculture, but have also played a major role in the development of biological control as a science. The crucial importance of sound systematics and biology, the ecological basis for natural enemy importation policy, methods of conserving and augmenting natural enemy populations, the effects of adverse factors on natural enemies and some methods for their mitigation, experimental check methods for evaluating the efficacy of natural enemies – these and other concepts, principles and methodologies of biological control have been established in various projects carried out on citrus (DeBach, 1969; Bennett *et al.*, 1976; Rosen and DeBach, 1978, 1979).

Of the various alternatives to chemical control available to us, biological control is by far the most successful, most promising and most desirable. It is an ideal form of pest management – inexpensive, often permanent, and decidedly non-hazardous (DeBach and Rosen, 1991). Unfortunately, however, it cannot be expected to provide an immediate solution to all pest problems in citriculture at the present time, simply because no effective natural enemies are currently available for certain 'key' pests, including, for instance, the Mediterranean fruit fly, *Ceratitis capitata*, in Israel, the citrus thrips, *Scirtothrips citri*, in California and the citrus leaf miner, *Phyllocnistis citrella*, in India. Until such enemies are discovered and put to use, chemical pesticides or some other means of control will continue to be required (Rosen, 1986, 1990).

Integrated pest management

Integrated pest management (IPM) provides a reasonable compromise, taking into account the desirability of biological control as well as the continuing need for some form of chemical control. IPM is the art of the possible. It represents a holistic approach, recognizing the unity of the agro-ecosystem and harmonizing all available methods to attain optimal pest control and environmental quality. In principle, effective IPM may be achieved by judicious use of relatively selective pesticides, only when absolutely necessary and in the least disruptive modes of application, in combination with a vigorous program of applied biological control. Other selective alternatives may be incorporated whenever applicable, but biological control by natural enemies should always be a major component (Rosen, 1986; DeBach and Rosen, 1991).

Substitution of broad-spectrum pesticides with selective ones, such as mineral oils or insect growth regulators, is an essential step towards the development of IPM. Adoption of selective modes of pesticide application, such as bait sprays, is another important step that should be taken whenever feasible. Most importantly, using chemical pesticides 'only when absolutely necessary' requires the establishment of reliable economic thresholds and accurate monitoring systems for all actual and potential pests occurring in the citrus agro-ecosystem. Other selective tactics – cultural, mechanical, physical and autocidal controls – would also tend to conserve existing natural enemies, and should be accompanied by importation of exotic species and, whenever necessary, also by augmentative releases of natural enemies. Special emphasis should be placed on inherently integrative approaches, such as breeding of pesticide-resistant natural enemies (e.g. Spollen *et al.*, 1994; Havron and Rosen, 1994).

Successful biological control projects have been the backbone of IPM programs on citrus worldwide. Following early successes in the control of the cottony cushion scale and citriculus mealybug by introduced natural enemies, the modern program in Israel began with the importation of *Aphytis holoxanthus* DeBach (Hymenoptera: Aphelinidae) against the Florida red scale in the 1950s, and modification of chemical control practices in order to conserve this effective natural enemy. Ultra-low-volume aerial bait sprays have been used against the Mediterranean fruit fly, and selective acaricides have replaced sulfur preparations for the control of mites. Further developments have included effective biological control of the Mediterranean black scale, the citrus whitefly and the bayberry whitefly, importation of natural enemies against several other pests, control of the citrus flower moth by pheromone mass-trapping, postharvest treatment of citrus fruit to remove scale insects and sooty mold by high-pressure rinsing, refinement of the supervised control system, and selection of *Aphytis* species for resistance to pesticides. As a result of these concerted efforts, a viable IPM program has been implemented, whereby only about 10% of the citrus acreage of Israel is currently still subject to non-selective pesticide applications (Rosen, 1980; Rössler and Rosen, 1990). Similar programs have been implemented in California, South Africa and elsewhere (DeBach and Rosen, 1991).

Success in IPM is obviously dependent on a thorough understanding of the citrus agro-ecosystem, including the biosystematics, biology and population dynamics of all pests and their natural enemies. Effective extension services are invaluable to the implementation of IPM.

Diseases

Citrus is subject to numerous diseases, some of which occur only in certain environments, while others, like *Phytophthora*, pose a serious problem in all citrus-growing areas. The common diseases, including postharvest fungal diseases, are listed in Table 5.11.

Fungal diseases

MAL SECCO

This disease is confined to the Mediterranean Basin, around the Black Sea and Asia Minor. It is most severe on lemon and citron, and also affects mandarins and their hybrids. The typical symptom in the canopy is veinal chlorosis. Leaves wilt and dry. Most infections occur through wounds in the branches and leaves. The fungus reaches the lumina of xylem vessels from which it spreads systemically and mostly upward. By peeling off the bark, red or orange coloration of the xylem is revealed. Infection may also occur via roots. A serological procedure for detection has been developed. The causal organism is *Phoma tracheiphila*. Black pycnidia with a neck are produced on branches under the epidermis. The optimum temperature for growth is around 20 °C. Infections occur in winter, as splashing raindrops disseminate the conidia. Some Italian lemon varieties are

Table 5.11 *Common diseases of citrus*

Common name	Causal organism
Fungal diseases in nurseries and orchards	
Alternaria brown spot of mandarins	*Alternaria citri*
Black spot	*Phyllostictina citricarpa*
Greasy spot	*Mycosphaerella horii*
	Mycosphaerella citri
Mal Secco	*Phoma tracheiphila*
Melanose	*Diaporthe citri*
Mushroom root-rot	*Armillaria mellea*
Phytophthora	*Phytophthora parasitica*
	Phytophthora citrophthora
Post bloom fruit drop disease	*Colletotrichum gloesporioides*
Scab	*Elsinoe fawcetii*
	Elsinoe australis
	Sphaceloma fawcetti var. *scabiosa*
Postharvest fungal diseases	
Blue mold	*Penicillium italicum*
Green mold	*Penicillium digitatum*
Diplodia stem end rot	*Physalospora rhodina*
Gray mold	*Botrytis cenerea*
Trichoderma rot	*Trichoderma viride*
Fusarium rot	*Fusarium moniliforme*
Anthracnose	*Colletotrichum gloesporioides*
Phomopsis stem end rot	*Diaporthe citri*
Sour rot	*Geotrichum candidum*
Alternaria rot	*Alternaria citri*
Brown rot	*Phytophthora citrophthora*
Bacterial diseases	
Blast and black pit	*Pseudomonas syringae*
Canker	*Xanthomonas campestris* cv. *citri*
Greening	Phloem limited bacterium
CVC	*Xylella fastidiosa*
Viruses and virus like agents	
Exocortis	Viroid-single standard RNA molecule, endocellular
Leprosis	Mite vectored bacilliform virus
Psorosis	Viruslike; not characterized in the past. Citrus ringspot virus (CRSV) implicated in Florida
Satsuma dwarf	An isometric virus, probably soil borne
Stubborn	Wall free mycoplasma, *Spiroplasma citri*
Tristeza	Closterovirus, flexuous, rod shaped particles
Miscellaneous	
Blight	Unknown

resistant, but the quality of their fruit is inferior. Vigorous growth and overhead irrigation accelerate development of the disease. Pruning out diseased shoots may limit the spread of the disease, as will windbreaks and antihail nets. Spraying the canopy from autumn until early spring with copper or benzimidazole fungicides protects the canopy from infection.

PHYTOPHTHORA-INDUCED DISEASES

Phytophthora spp. cause the most serious soilborne diseases. They are of worldwide distribution. Losses are heavy in nurseries (damping-off), in the orchard (foot rot gummosis) and on the fruits (brown rot).

Foot rot is caused by an injury to bark on the trunk or roots. Gummosis is the rotting of bark on the tree. The fungus grows in the cambium. Necrosis, commonly accompanied by abundant gum exudation, follows. Sometimes the gummosis occurs as foot rot, attacking the base of the trunk under the soil. *Phytophthora* can also cause a decay of feeder roots on susceptible rootstocks. Brown rot is a brown-colored decay of fruit, especially fruit near the ground that is splashed with soil. The most common and important *Phytophthora* spp. attacking citrus are *P. nicotiana* var. *parasitica* and *P. citrophthora*. *P. citrophthora* attacks aerial parts more frequently than *P. nicotiana* var. *parasitica*. Temperatures suitable for mycelial growth are lower for *P. citrophthora* (13–28°C) than for *P. parasitica* (30–32°C). *Phytophthora* spp. are endemic in the soil of orchards. Infection of suberized tree trunks requires wounds or cracks. Severe outbreaks usually follow periods of rainy weather. While most citrus scion cultivars are moderately to highly susceptible to bark infection, large differences in tolerance to *Phytophthora* are evident in rootstocks. Trifoliate orange (*Poncirus trifoliata*) is nearly immune. Citrumelo, and to a lesser extent Cleopatra, mandarin and sour orange are quite resistant. Sweet orange and some sources of Rough lemon are highly susceptible. Budding the trees well above the soil line, and maintaining the area under the tree dry and free of weeds and debris is of importance. Some chemicals and trunk sprays are used, mainly as complementary measures.

SCAB

Three different pathogens in the genera *Elsinoe* and *Sphaceloma* cause scab on citrus. The most widespread of these is *E. fawcetii*. Scab occurs in areas with summer rainfall; it has not been found in California and in most of the Mediterranean sea. Infection is followed by protuberances on the leaf and distortion of shoot apices. On very young fruit the infection causes warty outgrowths, with less raised pustules on grapefruit and sweet orange. During later infection, pustules are slightly raised and may coalesce into scabby areas and cause subsequent fruit cracking (about 2.5

hours of continuous wetting are required for infection). Fruit is susceptible up to about three months after petal fall. Fungicides are used both to
prevent production of conidia on scab pustules and also to provide
protectant action. Overhead sprinkling increases the risk of attacks.

ALTERNARIA (BROWN SPOT OF MANDARINS)

Some mandarins such as Dancy and Minneola tangelo are most susceptible to the attack of *Alternaria alternata* pathotype *citri*. Leaves and stem
show large necrotic blighted spots or smaller circular spots. Black spots
develop on the fruit a few days after infection. In early summer, lesions are
small and often a wound periderm is formed beneath the lesion. Later
infections cause black sunken spots. Leaves and stems show large necrotic
blighted spots or smaller circular spots. Infected leaves as well as early
infected fruit drop. The disease is more severe on vigorous trees, and with
overhead irrigation. Several sprays of fungicides are being used to combat
the disease; however, control is often difficult.

Bacterial diseases

CANKER

Canker disease is caused by several strains of the bacterium *Xanthomonas
campestris* pv. *citri*. The asiatic form, canker A, is the most widespread and
severe form. Canker originated in South East Asia. It occurs also in some
Pacific and Indian Ocean islands and in some South American countries.
Canker is spread by wind and rain. Frequent rainfalls during early shoot
growth and fruit development augment its severity. It causes leaf spots
and blemishes on the rind of the fruit. In severe cases, leaf and fruit drop
follow. Grapefruit and pummelo are highly susceptible, lemon and orange
moderately susceptible. On the leaves a yellow halo surrounds the lesions.
A more reliable symptom is the water-soaked margin which develops
around the necrotic tissue. The most critical period for infection is up to 90
days after petal fall. Rigid restrictions on the importation of propagating
material and of fruit from diseased areas have kept many areas free of the
disease. Windbreaks and nets are used in Japan to limit the severity of the
disease. Several copper sprays are given to protect the fruits, and, in some
countries, a prebloom spray is also applied.

GREENING

Greening is a very dangerous and highly destructive disease. It probably
originated in China and is now known to be caused by a phloem-limited
gram negative bacterium. The disease seriously affects yields in South
East Asia, India, the Philippines, Taiwan, Indonesia and Africa (except

the North African citrus-growing area). Trees infected when young often do not bear, while older bearing trees become nonproductive. The disease is called yellow shoot in China, leaf mottling in the Philippines and vein phloem degeneration in Indonesia.

The African form induces symptoms under cooler conditions than the Asian form (up to 32 °C). Most citrus cultivars, especially sweet oranges and mandarins, are severely affected. Leaves develop vein chlorosis, are small and symptoms recall zinc deficiency symptoms. Fruits on infected trees are small and poorly colored, hence the name of greening. The juice is low in sugar, and high in acids and bitterness. The causal agent is an endocellular gram negative bacterium with a three-layer envelope of cell wall, about 25 nm thick.

The greening agent can be graft transmitted, but transmission rates are irregular. Most natural spread occurs in late spring when vector populations are high. Two psyllids, *Trioza erytreae* (in Africa) and *Diaphorina citri* (in Asia), have been identified as vectors.

For identification, indexing is used, while electron microscopy and recently, DNA probes, are used for final positive identification. Gentisoyl glucoside, a fluorescent marker has been used for rapid detection.

When greening is endemic, control includes reduction of inoculum, and less often insecticides. Greening is suppressed by tree injection with tetracyclines. Budwood sources can be freed of greening by thermotherapy and shoot tip grafting.

A bacterial disease (CVC), which also causes limb dieback, has been identified in South America. It appears to be caused by the xylem-limited bacterium *Xylella fastidiosa* (Lee *et al.* 1991).

Virus and viruslike diseases

TRISTEZA

Citrus tristeza virus (CTV) is the most important virus pathogen of citrus and is a major problem affecting citrus production. Millions of trees on sour orange rootstock have been killed or abandoned because of CTV decline, in Brazil, USA, Spain, Argentina and elsewhere. Tristeza probably originated in Asia. Some isolates cause stem pitting of susceptible cultivars, even when propagated on tolerant rootstocks.

The main symptoms of the disease in commercial varieties are stunting, stem pitting, chlorosis and reduced fruit size. A symptom of great economic significance is decline, occurring mainly with sweet orange, mandarin and also grapefruit on sour orange rootstock. Some virus infected stock/scion combinations show phloem necrosis in the rootstock below the bud union. The girdling effect is followed by starch depletion in the rootstock and decline. In such trees, pinhole pitting often occurs at the inner face of the bark of the rootstocks. Severe stem pitting strains can seriously affect sweet oranges, grapefruit and mandarins on any rootstock. Some isolates cause stunting and chlorosis on sour orange, lemon and grapefruit seedlings. This reaction is called seedling yellows.

CTV is a member of the closterovirus group. The long flexuous particles contain single-stranded RNA with a molecular weight of about 6.5×10^6. The viral coat protein has a molecular weight of 26000. Biological identification of the disease is by grafting indicator seedlings of Mexican lime and observing for symptoms. Rapid identification can be achieved by the enzyme linked immunosorbent assay (ELISA). CTV is readily graft transmitted via a union between donor and receptor. Long distance spread of CTV has been due to propagation from infected buds. CTV is also transmitted by aphids. *Toxoptera citricida* is a most efficient vector, not yet present in North America and the Mediterranean. *Aphis gossypii* is a less efficient vector, and is less abundant on citrus than *T. citricida*. Additional vectors may also play a part in transmission. Natural spread has been slower in desert areas. CTV is not seed transmitted. Epidemics developed and caused considerable damage mainly on budded plants with sour orange as a rootstock, with vectors present and under conditions favorable for transmission. Quarantines, budwood certification and eradication have been implemented in many cases for the prevention and suppression of tristeza. Declining orchards are usually replanted on CTV-tolerant rootstocks. Employing such rootstocks will usually require the use of exocortis-free budwood, and may complicate the situation concerning blight, *Phytophthora* and lime induced chlorosis.

Control of CTV-induced stem pitting is difficult, especially when severe isolates are endemic. A solution adopted for protecting susceptible cultivars by deliberate infection with mild strains is employed on a large scale with Pera orange in Brazil, citrus in South Africa and grapefruit in Australia. Experiments are under way with coat protein mediated cross protection in order to protect susceptible combinations from severe decline. Coat protein mediated resistance (CP-MR) has been successfully employed in herbaceous crops. As several research groups have characterized different strains of CTV and the gene encoding the coat protein has been cloned and sequenced (Sekiya *et al.* 1991), it is expected that transgenic citrus plants expressing the CP gene of one or several viruses will be developed.

PSOROSIS

Psorosis is a complex of several diseases. The causal agent has not been identified but is presumed to be a virus. Recently, a two component RNA

virus, the citrus ringspot virus (CRSV) has been implicated in the etiology
of psorosis in *Florida*. Psorosis is present in old line trees in many citrus-
growing areas. The bark-scaling form causes tree weakening and decline.
Other forms, less detrimental, cause yield loss. The incidence of psorosis
has been much reduced by using virus-free budwood. Psorosis B, concave
gum, cristacortis and impietratura have somewhat similar symptoms to
psorosis, mainly in the leaves of certain sweet orange or sour orange
varieties. Symptoms are most evident in young leaves on spring or fall
flushes; they include chlorotic flecks vein banding and leaf mottling.
Scaling and flaking of the bark on trunks and limbs is also produced. The
disease agent is readily graft transmitted. Identification is made after graft
inoculation of sweet orange or mandarin seedlings. Certified budwood
and nucellar clones have been instrumental in eliminating psorosis.
Recently, thermotherapy and shoot tip grafting have been employed to
eliminate the disease from budwood sources.

STUBBORN

Stubborn was originally considered to be a virus, but it was found to be
caused by a mollicute named *Spiroplasma citri*. Stubborn is an important
disease in hot and arid growing areas, like California, the eastern Mediterranean, the Middle East and North Africa. It is called 'little leaf' in Israel.
Stubborn affects many hosts; all of them except those in Rutaceae and
Rosaceae are herbaceous. Oranges, grapefruit, mandarins are highly
sensitive. It is rarely lethal in citrus; however, young trees become
severely stunted, with shortened internodes, upright foliage, and cupped
and chlorotic or mottled leaves. Yield is adversely affected, the fruit is
small, lopsided and poorly colored. When mature trees are infected, the
symptoms are less striking.

Stubborn can be graft transmitted, but not all bud progeny from
infected trees manifest the disease. The disease is spread naturally by
leafhoppers. Apparently primary citrus infections result by dispersal of
vectors carrying *S. citri* from herbaceous hosts. Vector populations vary
from year to year. Natural spread is often rapid in young orchards. The
most rapid diagnostic method now used involves culturing on artificial
media and identification by microscopy or serology.

In areas where stubborn or the vectors are not endemic, control by using
disease-free budwood can be effective. In other cases, control is difficult.
Budwood sources can be rendered free of stubborn by shoot-tip grafting.

EXOCORTIS

The citrus exocortis viroid (CEV) causes tree stunting and bark scaling.
The viroid is readily graft transmitted, and has been disseminated by

using symptomless infected budwood. It is widespread in nearly all citrus-growing areas. Exocortis is mainly a disease of trees grafted on *Poncirus trifoliata*, trifoliate orange hybrids and Rangpur lime. Trees are stunted and yields are affected with often no appreciable effects on fruit quality. The above-mentioned sensitive rootstocks are all tristeza tolerant and employing CEV-free budwood enables their use. Recently, use of exocortis-infected budwood to induce tree dwarfing and early bearing has been also exploited.

CEV-infected sweet orange, grapefruit and mandarin are symptomless, but budded on a sensitive rootstock, show bark scaling and stunting. CEV is a circular, single-stranded RNA molecule with 371 nucleotides, highly base paired with a stable rod structure. Viroids smaller than CEV have been purified from Etrog citron, but they are not largely homologous with CEV.

CEV can be also transmitted mechanically by pruning tools. Seed transmission has not been confirmed. Indexing for CEV has been done by graft inoculation of Etrog citron. CEV can now be detected by identification via polyacrylamide electrophoresis of extracts and by hybridization with radioisotope-labeled nucleic acid probes.

While thermotherapy has not been effective in freeing citrus budwood from CEV infection, shoot-tip grafting has proven very effective.

Diseases of unknown causes

Blight, also known as YTD (young tree decline) and declinio (in Brazil) has been reported from Florida, Hawaii, Australia, Central and Latin America, Cuba and South Africa. It affects the water-uptake mechanisms of the host. Incidence of the disease is lower in mandarins and lemons than in grapefruit and sweet orange. Visual above-ground symptoms are nonspecific leaf wilt, stem dieback, water sprout production of the lower trunk and zinc deficiency pattern in the leaves. It has been known in Florida for nearly 100 years and losses there exceed half a million trees per year. Symptoms are not easily distinguished from other decline type diseases (CTV, greening, spreading decline). Weakened trees often succumb to infections by *Phytophthora* spp. All rootstocks are susceptible, but Rough lemon, Volkameriana, Rangpur and trifoliate orange are affected most and at an early stage. Blight can be diagnosed by water injection and zinc analysis in field trees. It has been root-graft transmitted from diseased to healthy trees (but there has been no transmission by grafting from the scion portion). No known biological agent has been definitely shown to be the transmissible causal agent.

Nematodes

Nematodes attacking citrus are small cylindrical nonsegmented worms, usually 0.4–1.0 mm long. The most widely occurring species is *Tylenchulus semipenetrans* Cobb. Another nematode of importance, causing decline, *Radopholus citrophilus*, has been reported only in Florida.

Low productivity of trees caused by large populations of the citrus nematode, *Tylenchulus semipenetrans*, feeding on the roots is often designated as 'slow decline'. Tree vigor is often reduced. Trees may also show leaf yellowing, sparse foliage and small fruit. The degree of damage caused depends on the soil, cultural practices and rootstocks. Soil samples are collected in places with many feeder roots and the female nematode population density counted. Sedentary females (0.35–0.4 mm long) are found on the surface of fibrous citrus roots under egg masses in a gelatinous matrix. They are obligate parasites and reproduction is facultatively parthenogenetic. So far, four biotypes of the citrus nematode have been identified. Population densities vary greatly from tree to tree and from season to season. Development of large populations is usually faster on fine textured or organic soils. Planting material free of the citrus nematode delays the development of damaging populations. Fumigation before planting has been helpful in reducing infestation of replants on old orchard sites. Rootstocks resistant to the citrus nematode all share *Poncirus trifoliata* germplasm. These include Carrizo, Troyer citranges, Swingle citrumelo and selected Poorman × *Poncirus* hybrids.

SPREADING DECLINE

Spreading decline is caused by the burrowing nematode *Radopholus citrophilus*, an obligate parasite with a wide host range. Spreading decline occurs only in central Florida on deep well-drained soil. Infested trees may have 50% fewer functional feeder roots. Under drought conditions trees may wilt. The number of infested trees increases with time, hence the name 'spreading decline'. Sampling for *R. citrophilus* is difficult because of the deep vertical distribution. Optimum temperature for nematode growth and reproduction is 24 °C. Of rootstocks tested, Milam (probably a rough lemon hybrid) and Carrizo citrange either tolerate the damage or reduce the nematode population. Some biotypes also affect these rootstocks.

Recommended reading

Baines, R. C., Van Gundy, S. D. and Du Charme, E. P. (1978).
   Nematodes attacking Citrus. In *The Citrus Industry*, Vol. IV, ed. W.

Reuther, E. C. Calavan, G. E. Carman, pp. 321–45. Berkeley: Division of Agricultural Sciences, University of California.

Bedford, E. C. G., ed. (1978). *Citrus Pests*. Science Bulletin 391, Department of Agricultural Technical Services, Republic of South Africa, 253 pp.

Castle, W. S. (1989). Citrus rootstocks. In *Rootstocks for Fruit Crops*, ed. R. C. Rom and R. F. Carlson, pp. 361–99. New York: John Wiley and Sons.

Chapman, H. D. (1968). The mineral nutrition of Citrus. In *The Citrus Industry*, Vol. II, ed. W. Reuther, L. D. Batchelor and H. J. Webber, pp. 127–289. Berkeley: Division of Agricultural Sciences, University of California.

Davis, F. S. and Albrigo, L. G. (1994). *Citrus*. Chapters 4 (Rootstocks) and 5 (Plant husbandry), pp. 83–107, 108–62. Wallingford, UK: CAB International.

DeBach, P. and Rosen, D. (1991). *Biological Control by Natural Enemies*, 2nd edn. Cambridge: Cambridge University Press. 440 pp.

Ebeling, W. (1959). *Subtropical Fruit Pests*. Berkeley: Division of Agricultural Sciences, University of California, 436 pp.

Embleton, T. W. and Reitz, H. J. (1973). Citrus fertilization. In *The Citrus Industry*, Vol. III, ed. W. Reuther, pp. 122–82. Berkeley: Division of Agricultural Sciences, University of California.

Embleton, T. W., Jones, W. W., Labanauskas, C. K. and Reuther, W. (1973). Leaf analysis as a diagnostic tool and guide to fertilization. In *The Citrus Industry*, Vol. III, ed. W. Reuther, pp. 183–210. Berkeley: Division of Agricultural Sciences, University of California.

Jones, H. G., Lakso, A. N. and Syversten, J. P. (1985). Physiological control of water status in temperate and subtropical trees. *Hort. Rev.*, 7: 301–44.

Klotz, K. J. (1978). Fungal, bacterial and nonparasitic diseases and injuries originating in the seedbed, nursery and orchard. In *The Citrus Industry*, Vol. IV, ed. W. Reuther, E. C. Calavan, G. E. Carman, pp. 1–66. Berkeley: Division of Agricultural Sciences, University of California.

Kriedemann, P. E. and Barrs, H. D. (1981). Citrus orchards. In *Water Deficits and Plant Growth*, Vol. VI, ed. T. T. Kozlowski, pp. 325–417. New York: Academic Press.

Marsh, A. W. (1973). Irrigation. In *The Citrus Industry*, Vol. III, ed. W. Reuther, pp. 230–79. Berkeley: Division of Agricultural Sciences, University of California.

Rieger, M. (1989). Freeze protection for horticultural crops. *Hort. Rev.*, 11: 45–109.

Wallace, J. M. (1978). Virus and viruslike diseases. In *The Citrus Industry*, Vol. IV, ed. W. Reuther, E. C. Calavan, G. E. Carman, pp. 67–184. Berkeley: Division of Agricultural Sciences, University of California.

Whiteside, J. O., Garnsey, S. M. and Timmer, L. W., eds. (1988). *Compendium of Citrus Diseases*. St. Paul, MN: American Phytopathological Society. 80 pp.

Wutscher, H. K. (1979). Citrus rootstocks. *Hort. Rev.*, 1: 237–69.

Yelenosky, G. (1985). Cold hardiness in Citrus. *Hort. Rev.*, 7: 201–38.

Literature cited

Alva, A. K. and Syvertsen, J. P. (1991). Soil and citrus tree nutrition are affected by salinized water. *Proc. Fla. State Hort. Soc.*, **104**: 135–8.

Anderson, J. A., Gusta, L. V., Buchanan, D. W. and Burke, M. J. (1983). Freezing of water in *Citrus* leaves. *J. Am. Soc. Hort. Sci.*, **108**: 397–400.

Arnon, D. I. and Stout, P. R. (1939). Molybdenum as essential element for higher plants. *Plant Physiol.* **14**: 599–602.

Ashizawa, M., Kondo, G. and Chujo, T. (1981). Effect of soil moisture on the daily change of fruit size in summer season of Satsuma mandarin. *Kagawa Univ. Faculty Agric. Tech. Bull.*, **32**: 87–94.

Augusti, M. and Almela, V. (1991). *Applicacion de fitoreguladores en citricultura.*, Spain: Aedis Editoral. 268 pp.

Avidov, Z. and Harpaz, I. (1969). *Plant Pests of Israel.* Jerusalem: Israel Universities Press. 549 pp.

Bar-Akiva, A. (1961). Biochemical indications as a means of distinguishing between iron and manganese deficiencies in citrus plants. *Nature*, **190**: 647–8.

Bar-Akiva, A. (1974). Nitrate estimation in citrus leaves as a means of evaluating nitrogen fertilizer requirements in citrus trees. *Proc. First Int. Congr. Citriculture*, Vol. 1, ed. O. Carpena, pp. 159–64. Murcia, Spain: Ministerio de Agricultura, Instituto Nacional de Investigaciones Agrarias.

Bar-Akiva, A. and Sternbaum, J. (1965). Possible use of the nitrate reductase activity of leaves as a measure on the nitrogen requirements of citrus trees. *Plant Cell Physiol.*, **6**: 575–7.

Barnette, R. M., De Busk, E. F., Hester, J. B. and Jones, H. W. (1931). The mineral analysis of a nineteen-year-old March Seedless grapefruit tree. *Citrus Industry*, **12**: 5–6, 34.

Bedford, E. C. G., ed. (1978). *Citrus Pests.* Science Bulletin 391, Department of Agricultural Technical Services, Republic of South Africa. 253 pp.

Bedford, E. C. G. (1990). Mechanical control: high-pressure rinsing of fruit. In *Armored Scale Insects, Their Biology, Natural Enemies and Control, World Crop Pests*, Vol. 4B, ed. D. Rosen, pp. 507–13. Amsterdam: Elsevier Science Publishers.

Bennett, F. D., Rosen, D., Cochereau, P. and Wood, B. J. (1976). Biological control of pests of tropical fruits and nuts. In *Theory and Practice of Biological Control*, ed. C. B. Huffaker and P. S. Messenger, pp. 359–65. New York: Academic Press.

Beutel, J. A. (1964). Soil moisture, weather and fruit growth. *Calif. Citrog.*, **49**: 372.

Beverly, R. B., Stark, J. C., Ojala, J. C. and Embleton, T. W. (1984). Nutrient diagnosis of 'Valencia' oranges by DRIS. *J. Am. Soc. Hort. Sci.*, **109**: 649–54.

Bielorai, H., Shalhevet, J. and Levy, Y. (1978). Grapefruit response to variable salinity in irrigation water and soil. *Irrig. Sci.*, **1**: 61–70.

Bielorai, H., Dasberg, S., Erner, Y. and Brum, M. (1982). The effect of various soil moisture regimes and fertilizers on citrus yield response under partial wetting of the root zone. In *Proc. Int. Soc. Citriculture, 1981*, ed. K. Matsumoto, pp. 585–9. Okitsu, Shizuoka, Japan: Okitsu Fruit Tree Research Station.

Bielorai, H., Dasberg, S., Erner, Y. and Brum, M. (1988). The effect of saline irrigation water on Shamouti orange production. *Proc. Sixth Int. Citrus Congr.*, Vol. II, ed. R. Goren and K. Mendel, pp. 707–15. Philadelphia/Rehovot: Balaban Publishers; Weikersheim, Germany: Margraf Scientific Books.

Bodenheimer, F. S. (1951). *Citrus Entomology in the Middle East.* The Hague: W. Junk. 663 pp.

Castle, W. C. (1980). Citrus root systems: their structure, functions, growth and relationship to tree performance. In *Proceedings Int. Soc. Citriculture, 1978*, ed. P. R. Cary, pp. 62–6. Griffith, NSW, Australia.

Chapman, H. D. (1968). The mineral nutrition of Citrus. In *The Citrus Industry*, Vol. II, ed. W. Reuther, L. D. Batchelor, H. J. Webber, pp. 127–89. Berkeley: Division of Agricultural Sciences, University of California.

Chapman, H. D. and Parker, E. R. (1942). Weekly absorption of nitrate by young, bearing orange trees growing out of doors in solution cultures. *Plant Physiol.*, **178**: 366–76.

Chapot, H. and Delucchi, V. L. (1964). *Maladies, troubles et ravageurs des agrumes au Maroc.* Rabat: Institut National de la Recherche Agronomique. 339 pp.

Coggins, C. W. Jr (1982). The influence of exogenous growth regulators on rind quality and internal quality of citrus fruits. In *Proc. Int. Soc. Citriculture*, Vol. 1, ed. K. Matsumoto, pp. 214–16. Okitsu, Shizuoka, Japan: Okitsu Fruit Tree Research Station.

Coggins, C. W. Jr and Hield, H. Z. (1968). Plant-growth regulators. In *The Citrus Industry*, Vol. II, ed. W. Reuther, D. L. Batchelor and H. J. Webber, pp. 371–89. Berkeley: Division of Agricultural Sciences, University of California.

Cohen, Y., Fuchs, M. and Green, G. L. (1981). Improvement of the heat pulse method for determining sap flow in trees. *Plant Cell Environ.*, **4**: 391–7.

Compere, H. (1961). The red scale and its insect enemies. *Hilgardia*, **31**: 173–278.

Cooper. W. C., Olsen, E. O., Maxwell, N. and Otey, G. (1956). Review of studies on adaptability of citrus varieties as rootstocks for grapefruit in Texas. *J. Rio Grande Val. Hort. Soc.*, **10**: 6–19.

Cruse, R. R., Wiegand, C. L. and Swanton, W. A. (1982). The effect of rainfall and irrigation management on citrus juice quality in Texas. *J. Am. Soc. Hortic. Sci.*, **197**: 767–70.

Cunningham, R. T. (1989). Male annihilation. In *Fruit Flies: Their Biology, Natural Enemies and Control, World Crop Pests*, Vol. 3B, ed. A. S. Robinson and G. Hooper, pp. 345–51. Amsterdam: Elsevier Science Publishers.

Darvas, B. and Varjas, L. (1990). Insect growth regulators. In *Armored Scale Insects: Their Biology, Natural Enemies and Control, World Crop Pests*, Vol. 4B, ed. D. Rosen, pp. 393–408. Amsterdam: Elsevier Science Publishers.

Davee, D. E., Righetti, T. L., Fallahi, E. and Robbins, S. (1986). An evaluation of the DRIS approach for identifying mineral limitations on yield in 'Napoleon' sweet cherry. *J. Am. Soc. Hort. Sci.*, **111**: 988–993.

Davies, F. S. (1986). Growth regulators improvement of postharvest quality. In *Fresh Citrus Fruits*, ed. W. F. Wardowski, S. Nagy and W. Grierson, pp. 79–99. Westport, CT: Avi Publishing.

Davis, R. S., Rippetoe, L. W. and Jackson, L. K. (1987). Intermittent microsprinkler irrigation and tree wraps for frost protection of young 'Hamlin' orange trees. *HortScience*, **22**: 206–8.

DeBach, P. (1969). Biological control of diaspine scale insects on citrus in California. In *Proc. 1st Int. Citrus Symp.* (Riverside, 1968) Vol. 2, pp. 801–15. Riverside: University of California.

DeBach, P. and Bartlett, B. (1951). Effects of insecticides on biological control of insect pests of citrus. *J. Econ. Entomol.*, **44**: 372–83.

DeBach, P. and Rosen, D. (1991). *Biological Control by Natural Enemies*, 2nd edn. Cambridge: Cambridge University Press. 440 pp.

DeBach, P., Rosen, D. and Kennett, C. E. (1971). Biological control of coccids by introduced natural enemies. In *Biological Control*, ed. C. B. Huffaker, pp. 165–94. New York: Plenum Press.

Du Plessis, S. F. and Koen, T. J. (1988). The effect on N and K fertilization on yield and fruit size of Valencia. In *Proc. Sixth Int. Citrus Congr.*, ed. R. Goren and K. Mendel, pp. 663–73. Philadelphia/Rehovot: Balaban Publishers; Weikersheim, Germany: Margraf Scientific Books.

Ebeling, W. (1959). *Subtropical Fruit Pests*. Berkeley: Division of Agricultural Sciences, University of California. 436 pp.

Ferrarius, G. B. (1646). Hesperides: *sive de malorum auerorum cultura et usu libri quatuor*. Rome: Hermanni Scheus. 480 pp.

Furr, J. R. and Taylor, C. A. (1939). Growth of lemon fruits in relation to moisture content of the soil. *USDA Tech. Bull.*, **640**.

Gausman, H. W. (1991). *Plant Biochemical Regulators*. New York: Marcel Dekker. 363 pp.

Girton, R. E. (1927). The growth of citrus seedlings as influenced by environmental factors. *Univ. Calif. Publ. Agr. Sci.*, **5**: 83–117.

Goldschmidt, E. E. and Greenberg, J. (1989). GA_3 on Citrus fruit surface: uptake and persistence. *Acta Hort.*, **239**, 55–61.

Greany, P. D. (1989). Host plant resistance to tephritids: an underexploited control strategy. In *Fruit Flies: Their Biology, Natural Enemies and Control, World Crop Pests*, Vol. 3A, ed. A. S. Robinson and G. Hopper, pp. 353–62. Amsterdam: Elsevier Science Publishers.

Greenberg, J. and Goldschmidt, E. E. (1990). Uptake of ^{14}C-gibberellic acid by mature grapefruit (*Citrus paradisi* Macf.) as affected by relative humidity and method of application. *Ann. Appl. Biol.*, **116**: 383–90.

Grieve, A. M. and Walker, R. A. (1983). Uptake and distribution of chloride, sodium and potassium ions in salt-treated Citrus plants. *Austr. J. Agric. Res.*, **34**: 133–43.

Harding, R. B., Miller, M. P. and Fireman, M. (1956). Sodium and chloride absorption by citrus leaves from sprinkler-applied water. *Citrus Leaves*, **36**: 6–8.

Harty, A. R. and van Staden, J. (1988). The use of growth retardants in citriculture. *Isr. J. Bot.*, **37**: 155–64.

Havron, A. and Rosen, D. (1994). Selection for pesticide resistance in two species of *Aphytis*. In *Advances in the Study of Aphytis (Hymenoptera: Aphelinidae)*, ed. D. Rosen, pp. 209–20. Andover, UK: Intercept.

Hilgeman, R. H. (1963). Trunk growth of 'Valencia' orange in relation to soil moisture and climate. *Proc. Am. Soc. Hort. Sci.*, **82**: 193–8.

Hilgeman, R. H. and Van Horn, C. W. (1954). Citrus growing in
    Arizona. *Ariz. Exp. Sta. Bull.*, **258**.
Hilgeman, R. H., Ehrler, C. E., Everling, C. E. and Sharp, F. O. (1969).
    Apparent transpiration and internal water stress in 'Valencia' oranges
    as affected by soil water, season and climate. *Proc. First Int. Citrus
    Symp.*, Vol. 3, ed. H. D. Chapman, pp. 1713–23. Riverside: University
    of California Press.
Hirose, K. (1982). Development of chemical thinners for commercial use
    for Satsuma mandarin in Japan. In *Proc. Int. Soc. Citriculture*, Vol. 1, ed.
    K. Matsumoto, pp. 256–60. Okitsu, Shizuoka, Japan: Okitsu Fruit
    Tree Research Station.
Jackson, L. (1991). *Citrus Growing in Florida.* Gainesville: University of
    Florida Press. 293 pp.
Jackson, J. L., Ayers, D. H. and Parsons, L. R. (1986). Peformance of
    individual covers for cold protection of young citrus. *Proc. Fla. State
    Hort. Soc.*, **99**: 18–23.
Jones, H. G., Lakso, A. N. and Syvertsen, J. P. (1985). Physiological
    control of water status in temperate and subtropical trees. *Hort. Rev.*, **7**:
    301–44.
Haufmann, M. R. (1968). Evaluation of the pressure chamber method
    for the measurement of water stress of citrus. *Proc. Am. Soc. Hort. Sci.*,
    93: 186–90.
Knapp, J. L. (ed.) (1987). *Florida Citrus Integrated Pest and Crop Management
    Handbook.* Gainesville: Florida Cooperative Extension Service, Institute
    of Food and Agricultural Sciences, University of Florida.
Knapp, J. L., Fasulo, T. R. and Brooks, R. F. (1987). Minor insect pests.
    In *Florida Citrus Integrated Pest and Crop Management Handbook*, ed. J. L.
    Knapp. Section XI, pp. XI1–110, Gainesville: Florida Cooperative
    Extension Service, Institute of Food and Agricultural Sciences,
    University of Florida.
Koo, R. C. J. (1979). The influence of N, K and irrigation on tree size
    and fruit production of 'Valencia' orange. *Proc. Fla. Hort. Soc.*, **92**: 10–
    13.
Koo, R. C. J. (1984). The importance of ground coverage by fertigation
    for citrus on sandy soils. *Fert. Issue*, **1**: 75–8.
Koo, R. C. J., Anderson, C. A., Callvert, D. A., Stewart, I., Tucker,
    D. P. H. and Wutscher, H. K. (1984). *Recommended Fertilizers and
    Nutritional Sprays for Citrus.* Bulletin 536-S University of Florida,
    Agricultural Experiment Station, Florida.
Kriedemann, P. E. and Barrs, H. D. (1981). Citrus orchards. In *Water
    Deficits and Plant Growth*, Vol. VI, ed. T. T. Kozlowski, pp. 325–417.
    New York: Academic Press.
Lee, R. F., Derrick, K. S., Baretta, M. J. G., Chagas, C. M. and Rosetti,
    V. (1991). Citrus variegated chlorosis, a new destructive disease of
    citrus in Brazil. *The Citrus Industry*, **72**: 12–13, 15.
Leonhardt, B., Rice, R. E., Harte, E. M. and Cunningham, R. T. (1984).
    Evaluation of dispensers containing trimedlure, the attractant for
    Mediterranean fruit fly (Diptera: Tephritidae). *J. Econ. Entomol.*, **77**:
    744–9.
Levitt, J. (1980). Water, salt, radiation and other stresses. In *Responses of
    Plants to Environmental Stresses*, Vol. II, ed. T. T. Kozlowski, pp. 365–
    488. New York: Academic Press.

Levy, Y., Bielorai, H. and Shalhevet, J. (1978). Long term effects on different irrigation regimes on grapefruit tree development and yield. *J. Am. Soc. Hort. Sci.*, **103**: 680–3.

Linares, F. and Valenzuela, R. (1993). Medfly program in Guatemala and Mexico: current situation. In *Fruit flies: biology and management*, ed. J. Aluja and P. Liedo, pp. 425–38. New York: Springer Verlag.

Lindow, S. E. (1982). Population dynamics of epiphytic ice nucleation active bacteria of frost sensitive plants and frost control by means of antagonistics bacteria. In *Plant Cold Hardiness and Freezing Stress: Mechanisma and Crop Implications*, Vol. 2, ed. P. H. Li and A. Sakai, pp. 395–416. New York: Academic Press.

Lombard, P. B., Stolzy, L. H., Garber, M. J. and Szuszkiewicz, B. J. (1965). Effect of climatic factors on fruit volume increase and leaf water deficits of citrus in relation to soil suction. *Soil Sci. Soc. Am. Proc.*, **29**: 205–8.

Loughridge, R. H. (1900). Effect of alkali on citrus trees. *Univ. Calif. Agr. Expt. Sta. Rpt.*, **1897–98**: 99–113.

Maas, E. V. and Hoffman, J. G. (1977). Crop salt tolerance – current assessment. *J. Irrig. Drain. Div. Am. Soc. Civ. Eng.*, **103**: (IR2) 115–34.

Marsh, A. W. (1973). Irrigation. In *The Citrus Industry*, Vol. III, ed. W. Reuther, pp. 230–79. Berkeley: Division of Agricultural Sciences, University of California.

Monselise, S. P. (1979). The use of growth regulators in citriculture: a review. *Sci. Hort.*, **11**: 151–62.

Mukai, T. and Kadoya, K. (1994). Citrus. In *Horticulture in Japan*, ed. K. Konishi, S. Iwahori, H. Kitagawa and T. Yakuwa, pp. 14–22. Tokyo: Publication Committee, XXIVth International Horticultural Congress, Asakura Publishing Co.

Nemec, S. (1978). Response of six citrus rootstocks to three species of Glomus, a mycorrhizal fungus. *Proc. Fla. State Hort. Soc.*, **91**: 10–14.

Nordby, H. E. and Yelenosky, G. (1982). Relationships of fatty acids to cold hardening of Citrus seedlings. *Plant Physiol.*, **70**: 132–95.

Nucifora, A. (1986). Cultural methods. In *Integrated Pests Control in Citrus-Groves*, ed. R. Cavalloro and E. di Martino, pp. 491–9. Rotterdam: A. A. Balkema.

Oppenheimer, H. R. and Elze, D. A. (1941). Irrigation of citrus according to physiological indicators. *Agric. Res. Stn. Rehovot Bull.*, **91**.

Orser, L., Staskowicz, B. J., Panopoulous, N. J., Dahlebeck, D. and Lindow, S. E. (1985). Cloning and ice nucleation genes in *Escherichia coli*. *J. Bacteriol*, **164**: 359–66.

Parsons, L. R., Wheaton, T. A. and Whitney, J. D. (1982). Undertree irrigation for cold protection with low volume microsprinklers. *HortScience*, **17**: 799–801.

Parsons, L. R., Wheaton, T. A. and Tucker, D. P. H. (1986). Florida freezes and the role of water in citrus cold protection. *HortScience*, **21**(1): cover page.

Pearson, G. A., Gross, J. A. and Hayward, H. E. (1957). The influence of salinity and water on growth and mineral composition of young grapefruit trees. *J. Am. Soc. Hort. Sci.*, **69**: 197–203.

Penman, H. L. (1948). Natural evaporation from open water, bare soil and grass. *Proc. R. Soc. London Ser A.*, **193**: 120–46.

Peynado, A. and Young, R. H. (1968). Moisture changes in intact citrus leaves monitored by a beta gauge technique. *Proc. Am. Soc. Hort. Sci.*, **92**: 211–20.

Purvis, A. C. (1980). Influence of canopy depth on susceptibility of 'Marsh' grapefruit to chilling injury. *HortScience*, **15**: 731–3.

Quayle, H. J. (1938a). *Insects of Citrus and Other Subtropical Fruits*. Ithaca, NY: Comstock. 583 pp.

Quayle, H. J. (1938b). The development of resistance to hydrocyanic acid in certain scale insects. *Hilgardia*, **11**: 183–210.

Ramos, C. and Kaufmann, M. R. (1979). Hydraulic resistance of rough lemon roots. *Physiol. Plant.*, **45**; 311–14.

Reuther, W. (1973). *The Citrus Industry*, Vol. III. Berkeley: University of California, Division of Agricultural Sciences, Publication 4014.

Reuther, W., Calavan, E. C. and Carman G. E. (eds.) (1989). *The Citrus Industry, Vol. 5, Crop Protection, Post Harvest Technology and Early History of Citrus Research in California*. University of California Division of Agriculture and Natural Resources. 374 pp.

Rieger, M. (1989). Freeze protection for horticultural crops. *Hort. Rev.*, **11**: 45–109.

Riehl, L. A. (1990). Control chemicals. In *Armored Scale Insects: Their Biology, Natural Enemies and Control*. World Crop Pests, Vol. 4B, ed. D. Rosen, pp. 365–92. Amsterdam: Elsevier Science Publishers.

Roessler, Y. (1989). Insecticidal bait and cover sprays. In *Fruit Flies: Their Biology, Natural Enemies and Control*. World Crop Pests, Vol. 3B, ed. A. S. Robinson and G. Hooper, pp. 329–36. Amsterdam: Elsevier Science Publishers.

Roose, M. L., Cheng, F. S. and Federici (1994). Origin, inheritance and effects of a dwarfing gene from the citrus rootstock *Poncirus trifoliata* 'Flying Dragon'. *HortScience*, **39**: 482 (Abstract).

Rosen, D. (1967a). Biological and integrated control of citrus pests in Israel. *J. Econ. Entomol.*, **60**: 1422–7.

Rosen, D. (1967b). Effect of commercial pesticides on the fecundity and survival of *Aphytis holoxanthus* (Hymenoptera: Aphelinidae). *Isr. J. Agr. Res.*, **17**: 47–52.

Rosen, D. (1980). Integrated control of citrus pests in Israel. In *Proceedings, International Symposium of IOBC/WPRS on Integrated Control in Agriculture and Forestry, 1979*, pp. 289–92. Vienna.

Rosen, D. (1986). Methodologies and strategies for pest control in citriculure. In *Integrated Pest Control in Citrus-Groves*, ed. R. Cavalloro and E. di Martino, pp. 521–30. Rotterdam: A. A. Balkema.

Rosen, D. (1990). IPM: background and basic methodology. In *Armored Scale Insects: Their Biology, Natural Enemies and Control*. World Crop Pests, Vol. 4B, ed. D. Rosen, pp. 515–17. Amsterdam: Elsevier Science Publishers.

Rosen, D. and DeBach, P. (1978). Diaspididae. In *Introduced Parasites and Predators of Arthropod Pests and Weeds: A World Review. Agricultural Handbook* 480, ed. C. P. Clausen, pp. 78–128. Washington DC: Agricultural Research Service, United States Department of Agriculture.

Rosen, D. and DeBach, P. (1979). *Species of* Aphytis *of the World*. Jerusalem: Israel Universities Press; The Hague: W. Junk. 801 pp.

Rössler, Y. and Rosen, D. (1990). A case history: IPM on citrus in Israel. In *Armored Scale Insects, Their Biology, Natural Enemies and Control*.

World Crop Pests, Vol. 4B, ed. D. Rosen, pp. 519–26. Amsterdam: Elsevier Science Publishers.

Sekiya, M. E., Lawrence, S. D., McCaffery, M. and Cline, K. (1991). Molecular cloning and nucleotide sequencing of the coat protein gene of citrus tristeza virus. *J. Gen. Virol.*, **72**: 1013–20.

Shalhevet, J. and Bielorai, H. (1978). Crop water requirement in relation to climate and soil. *Soil Sci.*, **125**: 240–7.

Shalhevet, J. and Levy, Y. (1990). Citrus trees. In *Irrigation of Agricultural Crops*, ed. B. A. Stewart and D. R. Nielson, pp. 951–86. Madison, WI: Agronomy Monograph No. 30, ASA-CSSA-SSSA.

Singh, S. P. and Rao, N. S. (1980). Relative susceptibilities of different species/varieties of citrus to leaf miner, *Phyllocnistis citrella* Stainton. In *Proceedings of the International Society of Citriculture, 1978*, ed. P. R. Cary, pp. 174–7. Griffith, NSW, Australia.

Smajtrla, A. G. (1993). Microirrigation for citrus production in Florida. *HortScience*, **28**, 295–8.

Spollen, K. M., Rosenheim, J. A. and Hoy, M. A. (1994). Intraspecific variation in response to pesticides in *Aphytis melinus* DeBach from California citrus: results of natural and artificial selection. In *Advances in the Study of Aphytis (Hymenoptera: Aphelinidae)*, ed. D. Rosen, pp. 190–208. Andover, UK: Intercept.

Stanhill, G. (1972). Recent development in water relation studies. Some examples from Israel citriculture. In *Proc. 18th Int. Hort. Congr.*, Vol. 4, ed. N. Goren and K. Mendel, pp. 367–79. Tel Aviv, Israel: ISHS and Ministry of Agriculture.

Sternlicht, M., Goldenberg, S., Nesbitt, B. F., Hall, D. R. and Lester, R. (1978). Field evaluation of the synthetic female sex pheromone of citrus flower moth, *Prays citri* (Mill.) (Lepidoptera: Yponomeutidae), and related compounds. *Phytoparasitica*, **6**: 101–13.

Swietlik and Faust, J. (1984). Foliar nutrition of fruit crops. *Hort. Rev.*, **6**: 287–355.

Syvertsen, J. P. (1981). Hydraulic conductivity of four commercial citrus rootstocks. *J. Am. Soc. Hort. Sci.*, **106**: 378–81.

Talhouk, A. S. (1975). Citrus pests throughout the world. In *Citrus Technical Monograph*, Vol. 4, ed. E. Hafliger, pp. 21–3. Basel: Ciba-Geigy Agro-chemicals.

Turrell, F. M. (1973). The science and technology of frost protection. In *The Citrus Industry*, Vol. III, ed. W. Reuther, pp. 338–446. Berkeley: Division of Agricultural Sciences, University of California.

Van Bavel, C. H. M., Newman, J. E. and Hilgeman, R. H. (1967). Climate and the estimated water use by an orange orchard. *Agric. Meteorol.*, **4**: 27–37.

Vardi, A., Spiegel-Roy, R., Ben-Hayyim, G., Neumann, H. and Shalhevet, J. (1988). Response of *Shamouti* orange and *Minneola* tangelo on six root-stocks to salt stress. In *Proc. Sixth Int. Citrus Congr.*, Vol. I, ed. R. Goren and K. Mendel, Philadelphia/Rehovot: Balaban Publishers. Weiker-sheim, Germany: Margraf Scientific Books.

Walker, R. R. (1986). Sodium exclusion and potassium-sodium selectivity in salt-treated trifoliate orange (*Poncirus trifoliata*) and Cleopatra mandarin (*Citrus reticulata*) plants. *Austr. J. Plant Physiol.*, **13**: 293–303.

Walker, R. R. and Douglas, T. J. (1983). Effect of salinity level on

uptake and distribution of chloride, sodium and potassium ions in
Citrus plants. *Austr. J. Agric. Res.*, **34**: 145–53.

Wilson, W. C. (1983). The use of exogenous plant growth regulators on
citrus. In *Plant Growth Regulating Chemicals*, Vol. 1, ed. L. G. Nickell,
pp. 207–32. Boca Raton, FL: CRC Press.

Wutscher, H. K. (1989). Alternation of fruit tree nutrition through
rootstocks. *HortScience*, **24**: 578–84.

Yankofsky, S. A., Levon, S., Bertold, T. and Sandlerman, H. (1981).
Some basic characteristics of bacterial freezing nuclei. *Am. Meteorol.
Soc.*, **20**: 1013–19.

Yelenosky, G. (1979). Water-stress induced cold hardening of young
citrus trees. *Proc. Am. Soc. Hort. Sci.*, **104**: 270–3.

Yelenosky, G. (1985). Cold hardiness in Citrus. *Hort. Rev.*, **7**: 201–38.

Young, R. H. (1969). Cold hardening in citrus seedlings as related to
artificial hardening conditions. *J. Am. Soc. Hort. Sci.*, **94**: 612–14.

Young, R. and Hearn, C. J. (1972). Screening citrus hybrids for cold
hardiness. *HortScience*, **7**: 14–18.

6

Genetic improvement in citrus

Introduction

MOST CITRUS CULTIVARS grown arose as chance seedlings or bud mutations of existing cultivars. A relatively small number of cultivars of widespread significance have so far originated from breeding programs. The genetics and breeding of citrus were reviewed by Cameron and Frost (1967), Cameron and Soost (1969), Soost and Cameron (1975), Vardi and Spiegel-Roy (1978) and more recently by Soost (1987) and Gmitter *et al.* (1992). Citrus cultivars are highly heterozygous (Soost and Cameron 1975). Little information has been obtained on the genetic control of traits. Leading cultivars represent subtle gene combinations, often of a highly elevated 'selection plateau'. Such combinations are disrupted by the sexual process. Many traits are polygenic as to their inheritance, being controlled by numerous genes. The probability of recombining genes in a successful hybrid to recreate the essential characters of a leading traditional cultivar is very low. Citrus breeding is also much hampered by the highly pronounced juvenility in both sexual and nucellar citrus seedlings. A further significant barrier to citrus hybridization and easy transfer of genetic material is the widely encountered apomixis (nucellar embryony) (Frost and Soost, 1968). The ability to cross within or between species in which few or no monoembryonic taxa are available – as in the orange and grapefruit – is highly restricted.

The situation has been somewhat alleviated by the increase in the list of monoembryonic cultivars produced by breeding; this may improve further by the addition of products of somatic hybridization as a result of protoplast fusion. Because of pronounced juvenility, often marked even during fruit-bearing age, breeding projects are long term, costly and require considerable field space. Seedless oranges, grapefruits and now also seedless mandarins (easy-peeling fruit) are an essential requirement for the fresh-fruit trade. Producing new seedless fruit cultivars is a stringent requirement for citrus breeders, complicated by the phenomenon of self-incompatibility, problems of sterility and the need for a pronounced parthenocarpic tendency. A further difficulty is the necessity of expanded tests, due to genotype × environment interaction and copious seed formation in the fruit under abundant cross-pollination in breeding and test plots. In addition, the rather strict adherence to fruit having exclusively orange and grapefruit genomes for industrial purposes, and the cultural and market establishment of proven, outstanding fresh-fruit cultivars such as Navel orange, Clementine, Marsh Seedless grapefruit present great difficulties for the initiation of the culture, expansion and trade of new cultivars. The tremendous impact of new mutants of the above-mentioned cultivars as well as those of Satsuma contributes to the trend of maintaining the present cultivars.

In spite of the difficulties and constraints enumerated, progress has been realized in the breeding of new easy-peeling cultivars, new rootstocks and in mutation breeding by irradiation of budwood and subsequent selection. Recent biotechnologies, protoplast fusion (Grosser and Gmitter, 1990), the production of transgenic citrus plants (Vardi *et al.*, 1990; Gmitter *et al.*, 1992; Hidaka and Omura, 1993) and their integration into citrus breeding programs and strategies offer new possibilities for genetic improvement in citrus.

Mutations and chimeras

Mutation, involving a change in DNA, is rather common in citrus; rates will vary according to cultivar. The observed frequency of mutation may also be influenced by environment, cultural practices such as pruning (resulting in the growth of buds that would otherwise remain latent), as well as by the type and number of trees that are being observed. Closer observation for changes in time of ripening, fruit color, seedlessness and possibly other characteristics can be more efficiently performed in smaller units of production (e.g. in Spain, Japan), especially if aided by guided, organized action (Satsuma in Japan). New mutants with valuable characteristics have been found and exploited, *inter alia*, in Navel orange, Marsh grapefruit, Clementine, Satsuma and several other 'Japanese' cultivars. Of particular interest are mutants for seedlessness, pigmented fruit, early and late ripening, and lower fruit acidity. Many unfavorable mutants exhibiting poor yield, abnormal fruit and atypical leaf characters have been disclosed by a wide survey started by A. D. Shamel in 1909, followed by a description of these types and observations of their performance by bud propagation (Shamel, 1943). Valuable spontaneous mutations have also been detected in nucellar seedlings, though it is often uncertain whether this reflects a change in chimeral status, a mutation, or a zygotic offspring.

A plant chimera is a combination of tissues of two or more genetic constitutions in the same plant or part of a plant. The term does not apply to originally budded and grafted plants, but to forms in which the combined genetic components grow side by side in the same part of the plant. Chimeras have been extensively studied and described by Neilson-Jones (1969) and Tilney-Bassett (1986). Synthetic chimeras formed by cutting off the grafted plants at the union have been described in Solanaceae (Winkler, 1908).

An account of similar synthetic chimeras in *Citrus* (called 'bizzarria') has been summarized by Strassburger (1907). Such a plant was described by Nati in 1624 (1929) (see Figure 6.1). Tanaka (1927) concluded that it has a core of citron and outer layers of sour orange. If one component forms an outer covering surrounding an inner core of another component the combination is defined as a periclinal chimera.

One growing-point cell layer or primary histogen (Dermen, 1945) may produce one or more layers of mature tissue. In the majority of

Figure 6.1   Bizzarria (a synthetic chimera) of citron and sour orange. After Tozzetti, from Nati (1929) edited by Ragioneri

dicotyledons there are two primary outer layers and one core layer. *Citrus*, too, has three histogenic layers.

Chimeras are of interest to the breeder and citrus grower. Useful characteristics will not be transmitted to the zygotic or nucellar progeny if a periclinal chimera is involved. Spiegel-Roy (1979) reported that some trees of Shamouti orange produce nucellar seedlings with Shamouti characters while others produce nucellar seedlings with characters of Beledi orange. Shamouti had been known to produce limbs with Beledi characteristics and its is assumed to have arisen around 1860 as a seedless mutation on a Beledi tree. It seems that Beledi tissue has since been lost in some sources of Shamouti, but not in others. A number of interesting chimeras have been described in Satsuma (Nishiura and Iwamasa, 1970).

The non recovery of low-acid individuals in progenies of the acidless orange when crossed with other acidless cultivars suggests a chimeral nature for acidless orange cultivars (Cameron and Soost, 1979). Another source of acidless orange seems to pass on the acidless character to the progeny (Barrett, 1982).

Color expression in parental clones and derived nucellar seedlings of grapefruit varieties also clearly indicate the often chimeral nature of pigmented variants and the fact that nucellar seedlings will maintain the color factor in the fruit provided layer L II producing the nucellus carries it (Cameron *et al.*, 1964; Cameron and Soost, 1969). Possible use of citrus fruit sector chimeras as a genetic resource for cultivar improvement has been pointed out by Iwamasa *et al.* (1978) and Bowman and Gmitter (1991).

Apomixis and nucellar polyembryony

Facultative apomixis by nucellar embryony, resulting in polyembryonic seed, largely complicates breeding efforts of citrus, apart from obscuring taxonomical relationships and sexual compatibilities. Controlled crossings using polyembryonic clones as female parents often yield few or no hybrid progeny. Facultative apomixis, especially when accompanied by sterility, and, especially in orange, by inbreeding depression, makes it very difficult to create large segregating, vigorous populations. The list of monoembryonic cultivars has increased recently. Somatic hybridization has the potential to create successfully tetraploid hybrids of polyembryonic parents; these are very difficult to produce sexually. Several approaches have dealt with the problem of how to increase and speed up the recovery of zygotic individuals in a mixed (zygotic and nucellar)

progeny. Techniques other than discriminating by morphological characters have been proposed to distinguish zygotic seedlings from nucellar
ones at an early stage. Attempts have also been made to find means to
reduce the number of nucellar embryos and increase the percentage of
survival and recovery of zygotic seedlings.

Techniques attempting to identify zygotic seedlings include thin-layer
chromatography of leaf flavonoids and coumarins (Tatum *et al.*, 1978),
root and leaf isozymes (Button *et al.*, 1976; Soost *et al.*, 1980; Roose and
Traugh, 1988), browning of shoot extracts (Geraci and Tusa, 1976; Esen
and Soost, 1978) and gas chromatography of leaf emissions (Weinbaum *et
al.*, 1982). Isozyme techniques make use of specific co-dominant alleles of
known inheritance. By identifying several heterozygous loci, the probability of a zygotic seedling carrying the same isozyme profile as the
maternal parent is very small (Torres *et al.*, 1982).

Reduction in the number of embryos per seed has been achieved by
high temperature treatment (Nakatani *et al.*, 1982) and by treatment of
flower buds with gamma rays (Ikeda, 1982). A graft-transmissible repres-
sor of nucellar embryogenesis has been reported in the citron (Soost,
1987). A significant increase in the percentage of zygotic seedlings in
Minneola tangelo (*C. paradisi* × *C. reticulata*) by treatment of young fruits
with 15 mg GA_3 ml^{-1} a month after anthesis has been reported (De Lange
and Vincent, 1978).

As it has been established (Kobayashi *et al.*, 1979) that cells developing
into nucellar embryos can be identified four to seven days before anthesis,
early intervention or application of compounds may ultimately be re-
warded. Another possible approach can be inferred from the findings
(Iwamasa *et al.*, 1970) that the zygotic embryo occupies a special apical
position.

Selection of nucellar seedlings (also called 'variants' or 'strains') from
established cultivars has been much practiced in the past and is also being
performed today. Many citrus clones were grown in the past from largely
apomictic seed, but this has ceded to the propagation of budded or grafted
trees, especially since the second half of the nineteenth century. Propagation from seed is still practiced in India with the Ponkan cultivar, and in
some other areas of South East Asia. Thorniness is very prominent in
nucellar seedlings and it may persist even in young trees of the second
budded generation from nucellar seedlings. Fruit characters are also often
inferior during the first years of fruiting, though favorable changes have
also been noted.

The most valuable horticultural characteristics of nucellar citrus selections have been high tree vigor, freedom from virus diseases and often
higher yields. Because of cases of later bearing and overly vigorous trees,

an alternative method of obtaining virus-free trees by shoot-tip grafting (Navarro *et al.*, 1975) is now being preferred in many cases to rejuvenation by the use of nucellar selections.

A high and consistent rate of nucellar embryony is required for citrus rootstocks, assuring a high degree of uniformity by seed propagation.

Mutation breeding (induced mutations)

Genetic improvement in *Citrus* by hybridization has been much hampered because of heterozygosity, reproduction by nucellar embryony and juvenility. Improvement in citrus has been largely the result of selection of naturally occurring somatic mutants. Many of the world's most important cultivars have arisen through somatic mutation. The citrus industry of the world is highly dependent on varieties such as Washington Navel, Valencia, Shamouti, Pera, Hamlin oranges, Marsh grapefruit, easy-peeling mandarins such as Satsuma, Clementine and of Eureka lemon – all of which are either completely seedless or very low seeded. Chromosome aberrations, resulting in more-or-less sterile gametes, are rather well tolerated. Backed by a highly developed parthenocarpic tendency and a vegetative mode of reproduction, aberrations are found in all of the above mentioned leading cultivars, with the exception of the self-incompatible Clementine, which has high pollen and ovule fertility. Mechanisms leading to seedlessness have been reviewed by Frost and Soost (1968) and Iwamasa (1966). A parthenocarpic tendency is an important prerequisite for seedlessness (Vardi and Spiegel-Roy, 1978, 1988). If the selection is of a pronounced parthenocarpic tendency and without an accompanying sizable decrease in fruit size, it may attain economic significance. It is therefore surprising that no more breeding efforts in *Citrus* have been devoted to the induction of mutants. However, the few achievements in this domain are striking and may, if continued, prove to be of considerable interest (Spiegel-Roy, 1990).

As to choice of methods employed in order to induce somatic variation in the clonally propagated fruit trees, positive results have been obtained almost exclusively by the use of irradiation. Variations in tree size, time of fruit maturity, seed number, as well as fruit color, have been induced. The subject has been reviewed by Lapins (1983), Broertjes and Van Harten (1988), and Spiegel-Roy (1990). In citrus, Hensz used X-rays and thermal neutrons on citrus seed and buds to induce somatic mutation. The 'Star Ruby' grapefruit was released as a seedless cultivar with improved color (originating from 'Hudson' with 40 seeds per fruit), following thermal neutron treatment of apomictic seed (Hensz, 1971). Recently 'Rio Red', with similar high internal color, was released; both irradiation and mutation played a part in giving rise to the new cultivar (Hensz, 1985). Radiosensitivity of different citrus species has been described (Spiegel-Roy and Vardi, 1989). Seedless forms of lemon, in one case from a cultivar bearing 25 seeds per fruit, have been obtained by budwood irradiation (Spiegel-Roy *et al.*, 1985, 1990). A seedless Minneola tangelo has also been produced (Spiegel-Roy and Vardi, 1989). Seedless mutants of orange and grapefruit were obtained in Florida (Hearn, 1984, 1986) and of orange in China (Zhou *et al.*, 1986). Changes in vegetative characteristics have also been induced by irradiation of lemon fruit with gamma rays 100–120 days after bloom. Culture of nucelli resulted in the isolation of two thornless mutants. Production of a low–medium acid, seedless, early-maturing mutant of Foster grapefruit from irradiated material has also been reported (Hearn, 1986; Gmitter *et al.*, 1992). Lowering acidity in existing cultivars could prove of great value in grapefruit, in grapefruit and pummelo hybrids, as well as in orange for industrial purposes.

Current irradiation techniques with fruit trees are based mainly on Zwintscher's technique, described by Lapins (1983), which aims to eliminate chimera formation by successive vegetatively propagated clonal generations from irradiated material.

Sterility as well as chromosome aberrations in natural seedless citrus cultivars have been described (Iwamasa, 1966; Frost and Soost, 1968). Anomalous chromosomal separation behaviour was also reported in two seedless mutants of sweet orange produced by irradiation of seed with Co^{60} gamma rays (Chen Shanchun *et al.*, 1991). During PMC meiosis, univalents and polyvalents have been observed at high frequencies in the two mutants. Assessing seed number provides a simple measure of estimating ovule fertility. Frost and Soost (1968) point out that while sterility in *Citrus* may be in some cases disadvantageous with respect to yield, it could also offset the effects of excessive fruit set, small fruit and alternate bearing.

Another possible advantage of induced and natural seedless mutants lies in the fact that while self-incompatible selections with parthenocarpic tendency will have numerous seeds when pollinated by fertile pollen from other clones, induced and natural mutants will most often combine pollen and ovule sterility and remain practically seedless under similar conditions.

Polyploidy

Polyploidy in *Citrus* has been extensively reviewed by Lee (1988). Chromosome counts performed in 33 genera of Rutaceae (Guerra, 1984a) including *Citrus* (8 species), *Poncirus trifoliata* and *Fortunella margarita*, have confirmed the number of chromosomes as 2n = 18. *Citrus sinensis* had a relatively small genome, 1C = 0.62 pg (Guerra, 1984b). Estimates of nuclear DNA by flow cytometry (Ollitrault and Michaux-Ferriere, 1994) yielded values of 0.8 pg to 1.0 pg in *Citrus*, with *Citrus medica* having the biggest nuclear genome and *Citrus reticulata* the smallest. While diploidy is the prevailing state in *Citrus*, tetraploid individuals have been widely reported since the apparently first report of Hong Kong wild kumquat, *Fortunella hindsii* (Champ.) Swingle, as a tetraploid. Most tetraploids were obtained as variant nucellar seedlings in seedling populations (Soost, 1987). Barrett and Hutchinson (1978) report a tetraploid frequency, depending on genotype, of less than 1% to 3%, and considerably higher in a few cases. The percentage found was higher in larger seeds, and has been shown to be influenced by environmental factors. Nucellar tetraploids can probably be recovered from most cultivars producing nucellar seedlings. Intentional production of tetraploids is of interest especially in mono-embryonic varieties, which do not produce nucellar offspring. Tachikawa (1971) and Barrett (1974) have reported on methods of applying colchicine for this purpose.

Interspecific somatic hybridization, as well as somatic hybridization by two polyembryonic parents, constitute a means of producing hetero-zygous tetraploid breeding parents (Grosser and Gmitter, 1990).

Tetraploids in citrus are characterized by slower growth, compact growth habit, typically broader, thicker, darker leaves and fruit with thicker rinds, less juice, (see Figure 6.2), larger oil glands and often lower fertility than corresponding diploids. Grapefruit tetraploids are comparatively more vigorous (Cameron and Soost, 1969). *Citrus* tetraploids have proven so far to be of practically no economic value, but of considerable interest in breeding for triploids.

Citrus triploids can be recognized morphologically, though less easily so than tetraploids. Leaves are thick, rounded and may be intermediate between those of their 2n and 4n parent. Many of the triploids are sterile, although fruitfulness in triploids has been found to be variable, but generally lower than in diploids. The number of seeds has been low and a reasonable parthenocarpic tendency in some of the triploids has been established. As low seed content is one of the main breeding objectives, selection of outstanding triploids is an important goal in citrus breeding. Two interesting triploid pummelo-grapefruits have been produced (Soost and Cameron, 1980, 1985). Because of nucellar embryony, tetraploids have been used as pollen parents in crosses, rather than the reciprocal 4n ♀ × 2n ♂ cross. Survival and recovery of triploids have been adversely affected by poor endosperm development and failure of embryo growth.

The ploidy ratio of endosperm to embryo is 5:3 instead of 3:2, as found in diploids.

Triploid seeds in *Citrus*, in contrast to those in apple, are smaller than diploid seeds of the same cross (Soost, 1987; Wakana *et al.*, 1982). By retaining only the small seeds, growing them *in vitro* on an agar medium and by early grafting of the seedlings on a rootstock, recovery of triploids is considerably enhanced. Endosperm culture has also enabled recovery of triploids (Wang and Chang, 1978; Gmitter *et al.*, 1990).

Production of triploid progeny will be further favored by use of mono-embryonic tetraploids as seed parents, as a result of application of improved methods for producing monoembryonic tetraploids with the aid of colchicine (Oiyama and Okudai, 1986).

Evaluation of populations from diploid monoembryonic seed parents has disclosed a significant proportion of triploids in the progeny, as a result of the production of diploid megagametophytes (Esen *et al.*, 1979; Geraci *et al.*, 1978). Triploids can thus also be recovered from diploid by diploid crosses, if the seed parent is producing a substantial percentage of diploid megagametophytes (Table 6.1). Esen *et al.* (1979) observed that diploid megagametophytes of Sukega develop after the first meiotic division, and can be thus considered as products of segregation and recombination.

Figure 6.2   Leaf and fruit of tetraploid 'Niva' (left) compared with clonal, diploid Niva (right)

Table 6.1 *Percentage of diploid megagametophytes among some citrus cultivars*

Cultivar	Diploid megagametophytes
Sukega	24.2
Temple	6.8
Clementine	1.0
King	7.0
Wilking	14.0
Fortune	20.0
Lisbon	1.0
Eureka	5.0
Poorman	0
Pummelo (CRC 2240)	0.5
Pummelo (CRC 2241)	0
Pummelo	0.1

Based on production of triploids in 2x × 2x or tetrapoids in 2x × 4x crosses.

Adapted from Soost (1987).

Hybridization

Problems due to heterozygosity, nucellar embryony and the prolonged juvenile period have been pointed out. Absolute or a high degree of gametic sterility is encountered in numerous citrus cultivars. The percentage of functional pollen varies among species and cultivars. Some of the most widely used commercial cultivars are deficient in this respect. Navel orange produces no viable pollen; Satsuma mandarin and Marsh grapefruit very little; lemons and most orange cultivars often have low amounts. Most cultivars of mandarin and pummelo produce largely functional pollen. Seedlessness and pollen sterility have been reviewed by Iwamasa (1966). Cultivars with a problem of non-functional pollen very often show comparable ovule abortion; though the pollen-sterile Washington Navel and, more so, Satsuma have some functional ovules. Degeneration before meiosis is also encountered.

In addition to absolute gametic sterility, self and, to some extent, cross incompatibility are also present in *Citrus*. Besides posing a problem for the breeder this also presents an opportunity for producing seedless cultivars, provided the parthenocarpic tendency is prominent and no opportunities for cross-pollination prevail in the citrus grove. Except for a publication by Soost (1969), little information is available on alleles and inheritance of self incompatibility in *Citrus*. Incompatibility is widespread in *C. grandis*. Self incompatibility has been also reported in *C. limon*, *C. limettoides* and some cultivars indigenous to Japan, in Clementine mandarin, and in Orlando and Minneola tangelo (both were derived from a cross between Duncan cv. (*C. paradisi*) and Dancy cv. (*C. reticulata*) and are cross incompatible). The list of self incompatible cultivars is on the increase, including hybrids between Clementine and Orlando. A report on unilateral cross incompatibility between *C. tachibana* Tan. and *C. hassaku* has been published by Ueno (1978). With the ancestry of many parents unknown, the presence of incompatibility in many progenies cannot be predicted. The grapefruit has probably inherited self incompatibility genes from the pummelo. Whether a similar situation occurs in some oranges is not known.

By utilizing self-incompatible parents, lack of fruiting may result in some individuals in the progeny. Self incompatibility will very often be obscured by sufficient fruit set in mixed hybrid selection plots. Hybrids of potential interest also have to be evaluated for fruitfulness in the absence of other pollen sources. The identification of suitable pollinating cultivars for new self-incompatible cultivars, which are unfruitful in the absence of cross pollination, is also required.

The period from seed to first fruiting is known as the juvenile period. Its length in citrus, often four to six years, is significant in both sexual seedlings (hybrids) and apomictic nucellar seedlings. Moreover, even with the advent of fruiting, some of its characteristics, such as thorniness and undesirable fruit shape, often persist. While there is a definite genetic component to the length of the juvenile period – oranges are slow to come into bearing compared with most mandarins and possibly also pummeloes, environmental conditions are also influential. However, the use of horticultural techniques such as girdling have not resulted in a significant shortening of the juvenile period. Fruiting is hastened, often by two years, by a labor-intensive method that is employed in Japan, entailing budding into older, grafted seedlings, training the scion to stakes and bending of shoots. Vardi and Spiegel-Roy (1988) have described the details of a method that has been used very effectively with hybrid progeny of Satsuma mandarin.

Extremely precocious flowering of grapefruit seedlings, notably Duncan, has been noted. Flowering (limited to one terminal flower) occurs after a half to one and a half years, but does not recur for the next four to five years. Iwamasa and Oba (1975) also obtained flowering in one pummelo cultivar, three tangelos (grapefruit × mandarin hybrids) and a hybrid between Satsuma and pummelo (*C. grandis*). Flowering occurred only in seedlings grown at maximum temperatures up to 20 °C, and minimum temperatures below 10 °C from November to March. The phenomenon seems mainly limited to grapefruit, grapefruit hybrids and pummelo hybrids. Only some of such seedlings have fertile pollen and ovules. A report on very early flowering of a *Poncirus trifoliata* seedling and its nucellar offspring has also been published (Yadav *et al.*, 1980). A specific antiserum reacting with an antigenic protein was recovered in larger quantities in tissues of mature grapefruit than in juvenile plants. The mature protein, with a molecular weight of *c.* 59.7 k Da was not found in other genera related to *Citrus*. Its content was higher in floral shoots of precociously flowering Marsh seedlings (Snowball *et al.*, 1991).

Distant hybridization

Many interspecific crosses have been performed in *Citrus*. Crosses with related genera, mainly *Poncirus*, have been made especially in rootstock studies and breeding (Swingle and Reece, 1967; Barrett, 1978). Group names indicating the parentage of interspecific and intergeneric hybrids have been noted in the literature. The most important of these include tangelo (tangerine × grapefruit), tangor (tangerine × orange), orangelo (orange × grapefruit) and citrange (*Poncirus* × sweet orange).

The scope of crosses between genera is on the increase, in an attempt to produce novel types of citrus rootstocks and cultivars, and for the future use of tetraploid products of somaclonal fusion.

Genetic aspects of citrus breeding: inbreeding and hybrid vigor

Great variability exists within and between *Citrus* species, for most characteristics of the tree and fruit. Variation is strongly expressed in hybrid progenies. Occasionally, hybrids exceed the limits of their parents in some character. In crosses with pummelo the range of progeny characters is usually closer to the pummelo parent. Interspecific crosses with *C. grandis* are usually of high vigor. Many weak hybrids have been obtained in narrow crosses within oranges. Inbred seedlings had much lower vigor than apomicts in *Poncirus* (Hirai *et al.*, 1986).

Mode of inheritance of characters

Characters suggestive of simple genetic control

Nucellar embryony seems to be inherited in a rather simple fashion. Hybrids from sexual monoembryonic parents are essentially mono-embryonic and considered homozygous for a recessive allele. In some cases a more complicated pattern of inheritance seems to be involved. The percentage of sexual hybrids in polyembryonic parents fluctuates from season to season and is strongly influenced by the pollen parent. The trifoliate leaf character of *Poncirus* is usually dominant over the monofoliolate leaf of *Citrus* (Cameron and Frost, 1968). However, segregation in some advanced generations does not confirm single gene inheritance. A cross of pummelo with citron yielded unilaterally 100% dwarfs; the reciprocal cross yielded only 50% dwarfs. With different pollen parents normal plants were obtained. Zygotic progeny of 'Flying Dragon', which is probably a mutation of *Poncirus trifoliata*, showed a 3:1 segregation for dwarfness, indicating a single dominant gene (Roose *et al.*, 1994). Curved thorns and twisted trunk were closely linked to, or pleiotropic effects of, the dwarfing gene. Three RAPD markers linked to the dwarfing gene were identified (Cheng and Roose, 1995). Extracts of young shoots show browning to be dominant to nonbrowning (Esen and Soost, 1978). In selfed Wilking 25% of the progeny show a lethal albino factor (Vardi and Spiegel-Roy, 1978). Inheritance of the bitter component flavanone neo-hesperidoside was found to be controlled by two dominant multiple genes (Matsumoto and Okudai, 1991).

Inheritance in a semi-qualitative manner has been indicated for acidity levels in the fruit as evidenced in crosses between acidless pummelo and other citrus species with medium acid fruit. Spiegel-Roy and Teich (1972) postulate that thorniness is controlled by more than one gene. More than one recessive gene seems to be involved in the inheritance of anthocyanin pigments (Vardi and Spiegel-Roy, 1978). Yellow rind color has been postulated to depend on two complementary recessive genes (Ligeng *et al.*, 1993). Quantitative range of character expression seems to exist for characters such as fruit size and probably also time of maturity, as well as for tolerance to *Phytophthora* (Hutchinson *et al.*, 1972).

The study of inheritance of parthenocarpic tendency and of seedlessness is of particular importance. Relatively simple inheritance may be indicated in some cases. Segregation for undeveloped anthers and sterile pollen versus normal anthers and fertile pollen tends to indicate a 1:1 pattern with Satsuma as seed parent (Iwamasa, 1967). Simply inherited, genetically controlled chromosome asynapis has been identified in selfed Wilking mandarin progeny (Vardi and Spiegel-Roy, 1982).

The recessive oligogenic character for seedlessness can perhaps also be inferred from mutation breeding results (Spiegel-Roy, 1990).

Biochemical and molecular markers

More than 20 isozyme loci, mostly highly polymorphic, have been genetically characterized in *Citrus*. Isozymes are being used for separation of zygotic seedlings, identification of somatic hybrids and studies in phylogeny. Visualization and interpretation of plant isoenzymes has been discussed by Wendel and Weeden (1989).

Analysis of RFLPs in a 'Clementine' derived population has shown most of the polymorphism to be due to insertions or deletions. A genomic library of 'Temple' and a cDNA library of *Poncirus* have been established as a source of probes (Gmitter *et al.*, 1992).

Construction of a genetic map of citrus nuclear genome, with isozymes and restriction fragment length polymorphisms (RFLPs) has been initiated (Jarrell *et al.*, 1992; Durham *et al.*, 1992). Genetic maps based on molecular markers could significantly aid early screening procedures by allowing selection to be based on phenotype as predicted by the genotype at molecular loci cosegregating with a specific phenotype.

Breeding aims

The main breeding aims for scion cultivars vary somewhat with different species and localities and in response to market trends. Cold tolerance is a foremost objective in certain environments. Attempts have been made to incorporate resistance to citrus tristeza virus (Barrett, 1990).

The main goals in *scion breeding* are the amount and regularity of the crop, fruit with good size, high quality, attractive appearance and color, very low seed content and easy peeling, season of ripening, high adaptation to maintenance of fruit on tree, transport and, in many cases storage.

Seedlessness is a prime requirement for fresh fruit. Otherwise high-quality cultivars or selections with seedy fruit are not accepted in many markets. The main pathways open at present for the breeder to obtain high-quality seedless selections are triploidy (though low fruit set is a problem with many triploids), use of highly parthenocarpic parents such as Satsuma, as well as exploitation of parthenocarpic tendencies in self-incompatible selections (Iwamasa and Oba, 1980). The asynapsis reported in inbred Wilking (Vardi and Spiegel-Roy, 1982) points to a further possibility for breeding seedless fruit. Of great significance is the observation, induction and exploitation of seedless variants in otherwise satisfactory, high-quality cultivars and selections.

A special objective of increasing significance is the breeding of grapefruit cultivars with low levels of acidity and, possibly, less bitterness. Another objective is the production of sweet oranges with high external and internal color, not necessarily due to anthocyanins. The extension of season of maturity by selecting fruit types maturing significantly earlier or later than existing cultivars is of great interest. This goal can be approached by hybridization as well as by a search for mutants, and possibly by mutation breeding.

Breeding for industrial purposes

Breeding citrus fruit specially adapted for industrial purposes has been performed mostly in Florida. The yield of total soluble solids per area planted and the juice content of the fruit are prime requirements, in addition to good color of the juice (though pigmented selections of grapefruit are less adapted) as well as lack of bitterness in the juice of sweet orange. Breeding for industrial purposes is, however, complicated by the demand for orange and grapefruit as the sole or nearly sole component for frozen concentrate. Some promising orange-like hybrids have been selected from breeding programs (Hearn, 1989; Spiegel-Roy and Vardi, 1987).

Cold hardiness

As large areas of citrus have low winter temperatures, several breeding programs aim to incorporate cold hardiness. This was also the first objective of USDA (United States Department of Agriculture) breeding work, initiated at the turn of the twentieth century. *Poncirus trifoliata* served as the main source for cold tolerance in crosses with *Citrus*. F_1 hybrids proved unsuitable as to fruit quality, but yielded at least two very valuable rootstocks (See Table 5.1). The program was restarted in 1973 (Barrett, 1982). *Eremocitrus glauca* and *Fortunella margarita* were also used as sources of cold hardiness. Fruits of hybrids with *Eremocitrus* have proved to be very acid. F_1 hybrids with *Poncirus* also had very acid fruit, often of acrid flavor. Selections with edible fruit quality, resembling orange, have been recovered from crosses of selected F_1 of *Poncirus trifoliata* × *C. paradisi* to *C. sinensis*. One of these, US 119, has been released (Barrett, 1990). It combines edibility with resistance to freezing and to citrus tristeza virus. High cold tolerance was found in open-pollinated progeny of pummelo–trifoliate hybrids (Yelenosky *et al.*, 1993). Cold hardiness is also a prime objective in Japan. *Poncirus*, *C. junos*, and Troyer citrange have proved to be the most cold hardy. A seedless selection named Kiyomi, from a Satsuma × Trovita orange cross, seems as cold hardy as Satsuma. Cold hardiness is a major breeding requirement for citrus in Russia. The main sources used have been *Poncirus* and Satsuma. *C. ichangensis* and *C. junos* have also been used, and some promising results have been reported (Soost, 1987). The main problem is the great difficulty in achieving satisfying fruit quality in progeny of crosses with *Poncirus* and in some other combinations. Changsha mandarin, Meyer lemon, Clementine and Natsudaidai (probably a pummelo hybrid) have also been employed in USA and Japan as sources of cold resistance. Further valuable sources of cold tolerance have been identified in China.

Disease and pest resistance

Sources of tolerance to diseases and pests in *Citrus* and *Citrus* relatives have been located. Their utilization in scion breeding programs as a prime objective is at present rather difficult because of the long juvenile period and the strict requirement as to fruit quality. The latter constraint does not apply in rootstock breeding.

Efforts are directed to achieving tolerance to 'Mal secco', caused by *Phoma tracheiphila*, in lemon and mandarin. So far, conventional breeding efforts have not been successful. Inheritance of resistance to the bacterial canker *Xanthomonas campestris* has been reported (Koizumi and Kuhara, 1982). Breeding aims may in the future incorporate resistance to diseases of fruit and foliage such as scab (*Elsinoe, Sphaceloma*) or Alternaria, which is particularly damaging to tangelos and some mandarins.

Goals of rootstock breeding

A range of rootstocks is available (Table 5.1). However, there is a need for further and better rootstocks as disease, virus and graft compatibility problems are present in many countries. Tolerance to cold is also of prime importance. Swingle citrumelo has been widely used recently, but compatibility problems limit its use to certain species and cultivars. Serious new rootstock-related diseases, such as blight, are also prominent in some areas. In certain environments only water of poorer quality is available, stressing the inadequacy of most present rootstocks under conditions of salinity and soil alkalinity. Size-controlling rootstocks which also induce scion precocity, which are firmly established in apple cultivars, are also needed in *Citrus*. One of the main aims of rootstock breeding is to find good substitutes for the sour orange as a rootstock. The latter cannot be used economically in many areas because of its susceptibility to the tristeza virus, but it has been outstanding in tolerance to calcareous and heavy soils, contributing also to the high quality of the fruit (high total soluble solids) of cultivars grafted onto it.

COLD TOLERANCE

Poncirus has been the main source of cold hardiness in rootstock breeding. Few hybrid rootstocks with *Poncirus* have equalled its cold hardiness (Yelenosky, 1985). Other breeding material of interest is: *C. junos, C. ichangensis,* Sunki mandarin, Shekwasha (*C. depressa* Hayata; probably a *C. tachibana* hybrid). Further sources of cold tolerance exist in genera other than *Citrus*. Somatic hybridization may promote their use.

SALT TOLERANCE

Rangpur lime (*C. limonia* Osbeck) and Cleopatra mandarins are considered to be salt-tolerant rootstocks. Reem and Furr (1976) have reported on breeding work for creating new salt-tolerant rootstocks. Some hybrids of Rangpur and Cleopatra with other *Citrus* cultivars or with *Poncirus* showed as little Cl^- accumulation in the scion as in Cleopatra mandarin. Vardi *et al.* (1988) reported on the relative tolerance of rootstocks grafted with Minneola tangelo and Shamouti orange. Rangpur and Cleopatra have been shown to be highly tolerant to Cl^- (Grieve and Walker, 1983; Sykes, 1985). *Poncirus* transported elevated Cl^-, but not Na^+ to the scion. In many cases, high Cl^- and Na^+ levels in tolerant selections may be carried to the scion, thus making them unsuited for imparting salt tolerance to grafted cultivars.

Excellent salt tolerance has been shown by *Severinia*. High salt tolerance has also been shown by Sunki mandarin, which is also tolerant to tristeza virus. Tolerance to calcareous soil is also of importance in certain areas. No rootstocks have been bred with special tolerance to high soil pH and/ or calcareous soil to replace the tolerant but CTV sensitive sour orange. Rangpur lime, Cleopatra and Sunki seem promising as parents in breeding for tolerance to highly calcareous soils. A recent report (Sudahono *et al.*, 1994) deals with the results of screening rootstocks for tolerance to bicarbonate-induced iron chlorosis.

PESTS, DISEASES, VIRUSES AND NEMATODES

Phytophthora parasitica and *P. citrophthora* are the causative agents of the most serious soilborne diseases in *Citrus*. Foot rot and root rot were responsible for the introduction of rootstocks and the use of grafted plants in *Citrus* during the mid 1800s (Castle, 1987). Rootstocks with good resistance are *Poncirus* and Swingle citrumelo (*C. paradisi* × *Poncirus*). *Poncirus* is widely used as a source of resistance in breeding programs (Soost, 1987; Spiegel-Roy *et al.*, 1988). Some resistance in *C. grandis* and *C. aurantifolia* and a higher degree in some *Citrus* relatives has been pointed out. Variable proportions of the progeny of crosses between *Poncirus* and susceptible parents show adequate tolerance. If high tolerance to

Phytophthora could be achieved by biotechnological methods, this would ease problems involved in breeding of rootstocks tolerant to tristeza virus and with other specific requirements, such as resistance to salt or blight.

A somewhat similar situation exists in imparting resistance against the citrus nematode, *Tylenchulus semipenetrans*, with *Poncirus* serving as the main source of resistance. Resistance of different sources of *Poncirus* and of Swingle citrumelo has been reported by Kaplan and O'Bannon (1981). Gottlieb *et al.* (1986) report on very high tolerance of selected hybrids of *Poncirus* with Poorman orange. Resistance to both *Tylenchulus semipenetrans* and the burrowing nematode, *Radopholus citrophilus* has not yet been accomplished. Milam rough lemon and Ridge Pineapple (*C. sinensis*) possess good tolerance to the burrowing nematode.

Tristeza virus, a closterovirus, is a major problem in most citrus producing areas, causing phloem necrosis below the graft union. The main source of resistance has been the immune *Poncirus*. Sweet orange, though tolerant, is most susceptible as a grafted plant on certain rootstocks to tristeza and is largely susceptible to *Phytophthora* on its own roots.

In the past, sweet orange scions were grafted on the test rootstock, inoculated and read for symptoms for 3–4 years. Enzyme linked immunosorbent assay (ELISA) has been helpful in determining the titre of tristeza in hybrids inoculated with the virus (Garnsey *et al.* 1981). The best hybrids so far have been crosses of *Poncirus* with sweet orange, grapefruit (Castle 1987) and lately also with Poorman orange (Spiegel-Roy *et al.*, 1988). Methods of genetic engineering are being employed in order to isolate a gene imparting resistance to the destructive virus. Dominant, single gene inheritance of resistance to CTV has been postulated (Barrett and Hutchinson, 1985). Resistance to citrus blight, a disease of yet unsolved etiology, is an important objective of programs attempting to develop rootstocks for humid citrus growing regions like Florida and Brazil. Sweet orange, sour orange and Cleopatra have proved to be more tolerant than *Poncirus*, Rough lemon and Rangpur.

Applications of tissue culture and biotechnology to the genetic improvement of citrus

Somatic embryogenesis

The culture of developing ovules and seeds at various periods before and after fertilization was initiated by Maheshwari and Rangaswamy (1958), concluding that only fertilized ovules respond in culture. Later, it was demonstrated that nucellar embryos can be obtained from nucellar explants of unfertilized ovules (Button and Bornman, 1971: Kochba *et al.*, 1972; Kochba and Spiegel-Roy, 1973). Embryogenic callus maintaining embryogenic competence for long periods was obtained from unfertilized ovules and nucelli (Kochba *et al.*, 1972).

Nucellar cells in *Citrus* are predetermined as embryogenic cells and growth substances are usually not required to obtain embryogenic callus, except in *C. limon*. Gibberellic acid was included in a medium for callus induction in Satsuma mandarin (Kunitake *et al.*, 1991). Embryogenic callus was also obtained using seeds from nearly mature fruit (Starrantino and Russo, 1980). *In vitro* systems in *Citrus* have been reviewed by Spiegel-Roy and Vardi (1984), Litz *et al.* (1985) and Vardi and Galun (1988). The effect of various sugars on embryogenesis and development has been studied (Kochba *et al.*, 1978; Kochba *et al.*, 1982). Galactose, lactose, raffinose and later also glycerol (Ben-Hayyim and Neumann, 1983) stimulated embryogenesis of nucellar callus. Only lactose was effective with Satsuma (Kunitake *et al.*, 1991).

SELECTIONS OF CALLUS LINES FOR TOLERANCE TO SALT, 2,4-D AND 'MAL SECCO' (*PHOMA TRACHEIPHILA*)

Salt-tolerant lines of Shamouti orange have been selected (Kochba *et al.*, 1982a) and plants tolerant to salt have been developed (Spiegel-Roy and Ben-Hayyim, 1985). Salt-tolerant lines of orange were produced after treatment of nucellar calli with gamma rays and EMS in China (Deng *et al.*, 1990). Selection for tolerance to 2,4-D did, however, impair embryogenic competence (Spiegel-Roy *et al.*, 1983).

Work by fusion of Femminello lemon and Valencia orange protoplasts has been initiated to produce genotypes imparting tolerance to *Phoma tracheiphila* (Tusa *et al.*, 1990).

The response of callus and protoplast to culture filtrate and to partially purified toxin of *Phoma tracheiphila* confirmed the sensitivity of 'Femminello' lemon, as contrasted to tolerance of 'Tarocco' orange to the 'mal secco' disease (Gentile *et al.*, 1992). Extracellular extracts of 'Femminello' nucellar callus selected for tolerance to *Phoma tracheiphila* have shown highly increased chitinase and glucanase activity (Gentile *et al.*, 1993).

ANDROGENESIS

In vitro differentiation of haploid plants by anther culture was first reported in *Poncirus trifoliata* (Hidaka *et al.*, 1979). Plantlet formation from anthers was also reported in *Citrus aurantium* (Hidaka *et al.*, 1982) and *Citrus sinensis* (Hidaka, 1984). The haploid nature of androgenic plants in

Citrus was, however, not clearly established. Their origin, as well as the conditions for the development of embryoids from microspores in the cases that have been reported, have been described by Hidaka and Omura (1989). Androgenesis in *Citrus* has been recently reviewed by Germana (1994).

CULTURE OF FURTHER *CITRUS* ORGANS

Stem and leaf explants have occasionally given rise to embryogenic callus (Grinblat, 1972; Chaturvedi and Mitra, 1974). Recently, juice vesicles of Satsuma mandarin yielded embryogenic callus (Nito and Iwamasa, 1990).

ENDOSPERM CULTURE

Triploid hybrid citrus plants were regenerated by somatic embryogenesis *in vitro* from endosperm-derived calli. Calli were induced from the cellular endosperm of *C. sinensis*, *C. paradisi* and *C. maxima* (excised 12–14 weeks post anthesis). Only sweet orange (*C. sinensis*) embryos developed and regenerated plants (Gmitter *et al.*, 1990). Triploid plantlets have also been previously reported from endosperm of *C. maxima* and *C. sinensis* (Wang and Chang, 1978).

SHOOT TIP GRAFTING

In vitro recovery of nucellar plants from monoembryonic cultivars of *C. reticulata* was first reported in 1969 (Rangan *et al.*, 1969). Plants of nucellar origin from the monoembryonic Clementine mandarin showed significant phenotypic variation (Juarez *et al.*, 1976).

The *in vitro* shoot-tip grafting method (Navarro *et al.*, 1975) provides a generally dependable procedure of obtaining true-to-type, virus-free material from established clones, including monoembryonic cultivars. Its developments and use have been reviewed by Navarro (1982).

Protoplast culture and regeneration

The protoplast to plant system in *Citrus* is highly efficient, reproducible and relatively simple. *Citrus* calli resulting from protoplast culture maintained on media devoid of growth substances have proven cytologically stable. Protoplasts were first cultured successfully (Vardi *et al.*, 1975) using embryogenic callus from ovule explants of Shamouti orange (*C. sinensis* Osb.) by Kochba *et al.* (1972). Ovule-derived nucellar callus of *Citrus* has been the generally appropriate source for protoplasts regenerating embryos and subsequently plants (Vardi *et al.*, 1982; Kobayashi *et al.*, 1983; Grosser *et al.*, 1990). Protoplast-derived citrus trees from Shamouti orange, Dancy and Ponkan mandarins, Villafranca lemon (Vardi *et al.*,

1982) and Trovita orange (Kobayashi, 1987) have set fruit remarkably similar to the fruit of the plants from which the protoplasts were originally derived (see Figure 6.3).

Recent advances in protoplast culture of citrus have been reviewed by Vardi and Galun (1988) and Grosser and Gmitter (1990).

Somatic embryogenesis of 'Femminello' lemon leaf-protoplast-derived cells, stimulated by co-culturing with embryogenic *C. sinensis* cells (Tusa *et al.*, 1990) has been postulated. Plants similar to grapefruit were recovered following fusion of *C. sinensis* protoplasts originating from embryogenic nucellar suspension culture with mesophyll-derived protoplasts of *C. paradisi* (Ohgawara *et al.*, 1989). Similarly, plantlets identical to Troyer citrange were regenerated from mesophyll protoplasts fused with nucellar cell suspension protoplasts of *C. sinensis* (Ohgawara *et al.*, 1991). The possibility of cybrid formation cannot be excluded and in some cases has been proven to occur. Callus recovered from proliferating zygotic embryos provided protoplasts, yielding embryos and subsequently trees of the monoembryonic *Microcitrus* (Vardi *et al.*, 1986). Plants have been regenerated from protoplasts originating from undeveloped ovules of mature fruit of Calamondin, *Citrus madurensis* Loureiro (Ling *et al.*, 1989).

Parasexual (somatic) hybrids

The successful establishment of plants from protoplasts has set the stage for somatic hybridization. The first example of successful somatic

Figure 6.3 Fruits of Dancy mandarin. C, from an orchard tree. P, from a protoplast derived tree. From Vardi and Galun (1988)

Table 6.2 *Intergeneric* Citrus *somatic hybrid plants produced from sexually compatible parents by protoplast fusion*

Parents and protoplast source[1]	Reference
C. sinensis 'Trovita' orange (ES)+ *Poncirus trifoliata*	Ohgawara *et al.* (1985)
C. sinensis 'Trovita' orange (ES)+ Troyer citrange (L) (*C. sinensis* × *P. trifoliata*)	Kobayashi & Ohgawara (1988)
C. sinensis 'Navel' orange + Troyer citrange	Ohgawara *et al.* (1991)
C. sinensis 'Hamlin' orange (ES)+ *P. trifoliata* Flying dragon (L)	Grosser *et al.* (1988b)
C. reticulata 'Cleopatra' mandarin (EC)+ *P. trifoliata* Flying Dragon (L)	Grosser *et al.* (1992)
C. reticulata 'Cleopatra' mandarin (EC)+ (*C. paradisi* × *P. trifoliata*) Swingle citrumelo	Grosser *et al.* (1992)
C. sinensis 'Valencia' orange (EC)+ *Fortunella crassifolia* 'Meiwa' kumquat (L)	Deng *et al.* (1992)
C. sinensis 'Valencia' orange (EC)+ (*C. sinensis* × *P. trifoliata*) Carrizo citrange (L)	Louzada *et al.* (1992)

[1] EC = embryogenic callus; ES = embryogenic suspension; L = leaf.
After Gmitter *et al.* (1992).

hybridization in *Citrus* was an intergeneric allotetraploid hybrid produced by fusion of embryogenic protoplasts derived from nucellar callus of Trovita orange (*Citrus sinensis*), fused with leaf protoplasts of *Poncirus trifoliata*, which lack the capacity for regeneration (Ohgawara *et al.*, 1985). Interspecific somatic hybrid plants were obtained from the fusion of Key lime (*Citrus aurantifolia*) and Valencia sweet orange (Grosser *et al.*, 1989). The number of somatic citrus hybrids is steadily increasing (Kobayashi and Ohgawara, 1988; Ohgawara and Kobayashi, 1991; Gmitter *et al.*, 1992; Fo *et al.*, 1994). The tetraploid state of fusion products has been verified using morphology, cytology, isozymes and DNA restriction patterns. Tables 6.2 and 6.3 list intergeneric somatic hybrids obtained from sexually compatible and incompatible parents. Protoplast fusion may prove an important means of bypassing the barriers of sexual hybridization, as demonstrated by fusion of protoplasts of Hamlin orange (*Citrus sinensis*), derived from an embryogenic suspension culture, with protoplasts from epicotyl-derived callus of *Severinia disticha* (Grosser *et al.*, 1988a).

Some of the intergeneric hybrids obtained collapsed after removal from the *in vitro* environment. This indicates that there are also limits to the genetic and physiological compatibilities with parasexual hybridization. New intraspecific hybrids in *Citrus* could prove to be of considerable

Table 6.3 *Intergeneric* Citrus *somatic hybrids produced from sexually incompatible parents by protoplast fusion*

Parents and protoplast source[1]	Reference
C. sinensis 'Hamlin' orange (ES) *Severinia disticha* (SDC)	Grosser *et al.* (1988a)
C. sinensis 'Hamlin' orange (ES) *Severinia buxifolia* (SDC)	Grosser *et al.* (1992)
C. reticulata 'Cleopatra' (ES) *Citropsis gilletiana* (L)	Grosser *et al.* (1990)
C. sinensis 'Hamlin' orange (EC) *Citropsis gilletinana* (SDS)	Grosser & Gmitter (1990)
C. sinensis 'Trovita' (EC) *Murraya paniculata* (L)	Shinozaki *et al.* (1992)
C. aurantifolia Mexican line (EC) *Feroniella lucida* (L)	Takayanagi *et al.* (1992)
C. aurantifolia Mexican line (EC) *Swinglea glutinosa* (L)	Takayanagi *et al.* (1992)

[1] EC = embryogenic callus; ES = embryogenic suspension; L = leaf; SDS = stem derived callus.

After Gmitter *et al.* (1992) and others.

importance (Gmitter *et al.*, 1992). An outline of the procedure followed for somatic hybridization is shown in Figure 6.4 (Kobayashi *et al.*, 1988a).

Possible applications of protoplast fusion are:

1 For rootstock improvement, through the production of allotetraploid hybrids, including the production of somatic hybrids between sexually incompatible genotypes. Their use, apart from horticultural characteristics, will depend also on their fertility.
2 In scion breeding, by using the new allotetraploid parents in crosses between tetraploids and diploids.

A more direct triploid production would result via gametic–somatic protoplast fusion. This could be achieved by successful isolation of haploid gametic protoplasts from flower buds, as accomplished in *Nicotiana* (Pirrie and Power, 1986) and *Petunia* (Lee and Power, 1988).

Somatic hybridization produces tetraploid hybrids possessing complementary traits from donor parents as exemplified by Kobayashi *et al.* (1988a). Successful recovery of diploid plants from tetraploid somatic hybrids – not yet accomplished – could greatly enhance the breeding of novel recombinant types.

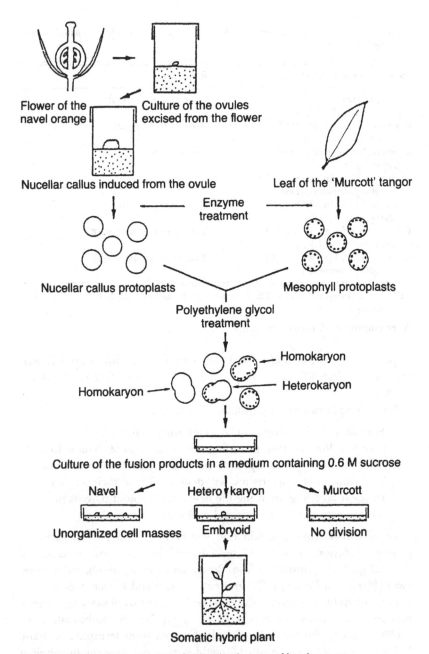

Figure 6.4 Model of somatic hybridization between Navel orange and Murcott tangor. After Kobayashi *et al.* (1988a)

Cybrids

The donor–recipient fusion method, involving inactivation of donor nuclei and metabolic inhibition of the recipient, followed by protoplast fusion (Galun and Aviv, 1988), has been successfully employed in producing cybrid *Citrus* plants with the nuclear coded morphology of the recipient fusion partner. Mitochondrial DNA restriction profiles (Vardi *et al.*, 1987) have disclosed novel mitochondrial genomes. The chloroplast genome of the wild *Citrus* relative *Microcitrus* has been transferred into cybrids possessing the nuclear genome of *Citrus* cultivars (Vardi *et al.*, 1989). Cybrids having the nuclear genome of a certain cultivar and alien organelles will aid breeding and genetic studies by providing information on cytoplasmic inheritance. Traits such as resistance to herbicides and to certain pathogens as well as cytoplasmic male sterility may turn out to be under control of organelle genomes in *Citrus*.

Transgenic plants in Citrus

A major attraction of genetic transformation is the possibility of adding a specific, advantageous trait to a cultivar or rootstock, thus avoiding sexual recombination, emergence of unfavorable characteristics and the need for lengthy backcross procedures.

Genetic transformation may allow the incorporation of heterologous genes into the genome. Genes isolated so far for insect, viral or herbicide resistance have been of bacterial or viral origin. Insertion of a viral coat protein gene into the genome of the species could protect against viral infection. Research on transformation of *Citrus* with the coat protein gene of CTV (Citrus tristeza virus) (see also Chapter 5) is being carried out in several laboratories.

Protoplasts isolated from suspensions of cultured cells of nucellar origin of orange (*Citrus sinensis*) were transformed by direct DNA transfer (Kobayashi and Uchimiya, 1989). Protoplasts were treated with a bacterial plasmic DNA carrying a chimeric gene. Transformation frequency was in the order of 10^{-6}. Transgenic plants were obtained by Vardi *et al.* (1990). Plasmid pCAP 212 DNA harboring the coding sequences of neomycin phosphotransferase and chloramphenicol acetyltransferase genes were introduced into *Citrus jambhiri* protoplasts through polyethyleneglycol treatment. Microcolonies were selected on an agar medium containing paramomycin sulfate. Transgenic plants were regenerated from two of the nine stably transformed embryogenic clones. Their transgenic nature was verified by neomycin phosphotransferase activity or Southern hybridization.

Hidaka *et al.* (1990) co-cultured a callus line of Washington navel orange and pollen embryoid callus of Trovita orange with strains of *Agrobacterium tumefaciens* harboring Npt II or a hygromycin phosphotransferase gene. Callus colonies formed on media with kanamycin or hygromycin embryoids. Subsequently, plantlets were formed and transformation of calli and of one plantlet was confirmed. Hidaka and Omura (1993) also report on transformation experiments comparing use of direct DNA by electroporation and *Agrobacterium* mediated transformation. An improved promoter gave a five-fold increase in β-glucoronidase (GUS), a marker visualized by a simple histochemical assay (Jefferson *et al.*, 1987). GUS activity was highest in leaf wings of transformed sweet orange plants.

Moore *et al.* (1989) used explants of internodal stem segments from *in vitro* germinated seedlings, mostly of Carrizo (*C. sinensis* × *Poncirus trifoliata*) and also of Swingle citrumelo (*C. paradisi* × *Poncirus trifoliata*) and Key lime (*C. aurantifolia*). These were co-cultured with *Agrobacterium tumefaciens* harboring the binary vector pMON9793 containing a chimeric gene for the expression of the NPT II coding sequence and a gene for the GUS marker. The antibiotic Mefoxin (cefotixin, Merck Sharp and Dohme) was shown to inhibit bacterial growth but not plant regeneration (Gmitter *et al.*, 1992). Five per cent of the shoots of each genotype, regenerated in the presence of 100 mg/ml kanamycin, were GUS positive. Two Carrizo plants were confirmed to be transgenic (Moore *et al.*, 1992).

The introduction of alien DNA with drug resistance markers opens possibilities for the development of novel transgenic citrus with advantageous breeding traits.

While improvements in transformation techniques and especially frequency are still required, the identification, isolation and cloning of economically significant genes are essential for the exploitation of the ability to create transgenic plants. Isolation and construction of genes controlling seed production in *Citrus*, either by preventing early seed development or by inhibiting pollen formation, are also being investigated (Koltunow, 1993).

Recommended reading

Cameron, J. W. and Frost, H. B. (1968). Genetics, breeding and nucellar embryony. In *The Citrus Industry*, Vol II, ed. W. Reuther, L. D. Batchelor and H. J. Webber, pp. 325–81. Berkeley: Division of Agricultural Sciences, University of California.

Gmitter, F. G. Jr., Grosser, J. W. and Moore, G. A. (1992). Citrus. In *Biotechnology of Perennial Fruit Crops*, ed. F. A. Hammerschlag and R. E. Litz, pp. 335–69. Wallingford, Oxon: CAB International.

Grosser, J. W. and Gmitter, F. G. Jr (1990). Protoplast fusion and citrus improvement. *Plant Breed. Rev.* **8**: 339–74.

Soost, R. K. (1987). Breeding, citrus genetics and nucellar embryony. In *Improving Vegetatively Propagated Crops*, ed. A. J. Abbott and R. K. Atkin, pp. 83–110. London: Academic Press.

Soost, R. K. and Cameron, J. W. (1975). Citrus. In *Advances in Fruit Breeding*, ed. J. Janick and J. N. Moore, pp. 507–40. West Lafayette, Indiana: Purdue University Press.

Spiegel-Roy, P. and Vardi, A. (1984). *Citrus*. In *Handbook of Plant Cell Culture*, Vol. 3, ed. P. V. Ammirato, D. A. Evans, W. R. Sharp, and Y. Yamada, pp. 355–72. New York, London: Macmillan.

Literature cited

Barrett, H. C. (1974). Colchicine-induced polyploidy in *Citrus*. *Bot. Gaz.*, **135**: 29–34.

Barrett, H. C. (1978). Intergeneric hybridization of *Citrus* and other genera in citrus improvement. In *Proc. Int. Soc. Citriculture, 1977*, ed. W. Grierson, pp. 586–9. Lake Alfred, FL: ISC.

Barrett, H. C. (1982). Breeding cold-hardy citrus scion cultivars. In *Proc. Int. Soc. Citriculture, 1981*, ed. K. Matsumoto, pp. 61–6. Okitsu, Shizuoka, Japan: Okitsu Fruit Tree Research Station.

Barrett, H. C. (1990). US 119, an intergeneric hybrid citrus scion breeding line. *HortScience*, **25**: 1670–1.

Barrett, H. C. and Hutchinson, D. J. (1978). Spontaneous tetraploidy in apomictic seedlings of *Citrus*. *Econ. Bot.*, **32**: 27–45.

Barrett, H. C. and Hutchinson, D. J. (1985). Rootstock development, screening and selection for disease tolerance and horticultural characteristics. *Fruit Var. J.*, **39**: 21–5.

Ben-Hayyim, G. and Neumann, H. (1983). Stimulatory effect of glycerol on growth and somatic embryogenesis in *Citrus* callus culture. *Z. Pflanzenphysiol.*, **110**: 331–7.

Bowman, Kim D. and Gmitter, F. G. Jr (1991). Citrus fruit sector chimeras as a genetic resource for cultivar improvement. *J. Am. Soc. Hort. Sci.* **116**: 888–93.

Broertjes, C. and Van Harten, A. M. (1988). *Applied Mutation Breeding for Vegetatively Propagated Crops. Developments in Crop Science*. Vol. 12, Oxford: Elsevier, 345 pp.

Button, J. and Bornman, C. H. (1971). Development of nucellar plants from unpollinated and unfertilized ovules of the Washington Navel orange *in vitro*. *J. S. Afr. Bot.* **37**: 127–34.

Button, J., Vardi, A. and Spiegel-Roy, P. (1976). Root peroxidase as an aid in *Citrus* breeding and taxonomy. *Theor. Appl. Genet.*, **47**: 119–23.

Cameron, J. W. and Frost, H. B. (1968). Genetics, breeding and nucellar embryony. In *The Citrus Industry*, Vol. II, ed. W. Reuther, L. D. Batchelor and H. J. Webber, pp. 325–81. Berkeley: Division of Agricultural Sciences, University of California.

Cameron, J. W. and Soost, R. K. (1969). Citrus. In *Outlines of Perennial Crop Breeding in the Tropics*, ed. F. R. Ferwerda and F. Wit, pp. 129–62. Wageningen: Veenman and Zonen.

Cameron, J. W. and Soost, R. K. (1979). Absence of acidless progeny from crosses of acidless × acidless *Citrus* cultivars. *J. Am. Soc. Hort. Sci.*, **104**: 220–2.

Cameron, J. W., Soost, R. K. and Olson, E. D. (1964). Chimeral basis for color in pink and red grapefruit. *J. Hered.*, **55**: 23–8.

Castle, W. S. (1987). Citrus rootstocks. In *Rootstocks for Fruit Crops*, ed. R. C. Rom and R. F. Carlson, pp. 361–99. New York: J. Wiley and Sons.

Chaturvedi, H. C. and Mitra, G. C. (1974). Clonal propagation of *Citrus* from somatic callus culture. *HortScience*, **9**: 118–20.

Chen Shanchun, Gao Feng and Zhang Jinren (1991). Studies on the seedless character of citrus induced by irradiation. *Mutation Breeding Newsl.*, **37**: 8–9.

Cheng, F. S. and Roose, M. L. (1995). Origin and inheritance of dwarfing by the *Citrus* rootstock *Poncirus trifoliata* 'Flying Dragon'. *J. Am. Soc. Hort. Sci.*, **120**: 286–91.

DeLange, J. H. and Vincent, A. P. (1978). Citrus breeding: new techniques in stimulation of hybrid production and identification of zygotic embryos and seedlings. In *Proc. Int. Soc. Citriculture, 1977*, ed. W. Grierson, pp. 589–95. Lake Alfred. FL: ISC.

Deng, Z., Zhang, W. and Wan, S. (1990). In vitro induction, biochemical analysis and protoplast plant regeneration from NaCl tolerant lines in Citrus. In *Proc. Int. Citrus Symp.* Guangzhou, China 1990, ed. B. Huang and Q. Yang, pp. 236–70. Beijing: International Academic Publishers. (In English.)

Deng, X. X., Grosser, J. W. and Gmitter, F. G. Jr (1992). Intergeneric somatic hybrid plants from protoplast fusion of *Fortunella crassifolia* 'Meiwa' with *Citrus sinensis* 'Valencia'. *Sci. Hort.*, **49**: 55–62.

Dermen, H. (1945). The mechanism of colchicine induced cyto-histological changes in cranberry. *Am. J. Bot.*, **32**: 387–94.

Durham, R. E., Liou, P. C., Gmitter, F. G. Jr and Moore, G. A. (1992). Linkage of fragment length polymorphism and isozymes in *Citrus*. *Theor. Appl. Genet.*, **84**: 39–48.

Esen, A. and Soost, R. K. (1978). Separation of nucellar and zygotic citrus seedlings by use of polyphenol-oxidase-catalyzed browning. In *Proc. Int. Soc. Citriculture, 1977*, ed. W. Grierson, pp. 616–18. Lake Alfred, FL: ISC.

Esen, A., Soost, R. K. and Geraci, G. (1979). Genetic evidence for the origin of diploid megagametophytes in *Citrus*. *J. Hered.*, **70**: 5–8.

Fo, F. A. A. M., Grosser, J. W. and Gmitter, F. G. Jr (1994). Production of seven new intergeneric hybrids for citrus rootstock improvement. *HortScience*, **29**: 482 (Abstract no. 362).

Frost, H. B. and Soost, R. K. (1968). Seed reproduction: development of gametes and embryos. In *The Citrus Industry*, Vol. II, ed. W. Reuther, L. D. Batchelor and H. J. Webber, pp. 292–334. Berkeley: Division of Agricultural Sciences, University of California.

Galun, E. and Aviv, D. (1988). Organelle transfer. In *Plant Molecular Biology. Methods in Enzymology*, Vol. 118, ed. A. Weissbach and H. Weissbach, pp. 595–611. Orlando, FL: Academic Press.

Garnsey, S., Barrett, H. and Hutchison, D. (1981). Resistance to citrus tristeza virus in citrus hybrids as determined by enzyme linked immunosorbent assay. *Phytopathology*, **71**: 875.

Gentile, A., Tribulato, E., Continella, G. and Vardi, A. (1992). Differential response of citrus calli and protoplasts to culture filtrate and toxin of *Phoma tracheophila*. *Theor. Appl. Genet.*, **83**: 759–64.

Gentile, A., Tribulato, E., Deng, Z. N., Galun, E., Fluhr, R. and Vardi, A. (1993). Nucellar callus of 'Femminello' lemon selected for tolerance to *Phoma tracheiphila* toxin, shows enhanced release of chitinase and glucanase into the culture medium. *Theor. Appl. Genet.*, **86**: 527–32.

Geraci, G. and Tusa, N. (1976). Distinzione tra semenzali nucellari e zigotici de arancio amaro per mezzo di testi biochimici. *Riv. Ortoflorofrutticoltura. Ital.*, **60**: 27–32. (Italian, with English summary.)

Geraci, G., De Pasquale, F. and Tusa, N. (1978). Percentages of spontaneous triploids in progenies of diploid lemons and mandarins. In *Proc. Int. Soc. Citriculture, 1977*, ed. W. Grierson, pp. 596–7. Lake Alfred, FL: ISC.

Germana, M. A. (1994). Androgenesis in Citrus: a review. In *Proc. Int. Soc. Citriculture, 1992*, Vol 1, ed. E. Tribulato, A. Gentile and G. Reforgiato, pp. 183–9. Catania, Italy: MSC Congress.

Gmitter, F. G. Jr., Ling, X. B. and Deng X. X. (1990). Induction of triploid *Citrus* plants from endosperm calli *in vitro*. *Theor. Appl. Genet.*, **80**: 785–90.

Gmitter, F. G., Jr., Grosser, J. W. and Moore, G. A. (1992). *Citrus*. In *Biotechnology of Perennial Fruit Crops*, ed. F. A. Hammerschlag and R. E. Litz, pp. 335–69. Wallingford, Oxon, UK: CAB International.

Gottlieb, Y., Cohen, E. and Spiegel-Roy, P. (1986). Biotypes of the citrus nematode (*Tylenchulus semipenetrans* Cobb) in Israel. *Phytoparasitica*, **14**: 193–8.

Grieve, A. M. and Walker, R. A. (1983). Uptake and distribution of chloride, sodium and potassium ions in salt treated citrus plants. *Austr. J. Agric. Res.*, **34**: 113–43.

Grinblat, U. (1972). Differentiation of citrus stem *in vitro*. *J. Am. Soc. Hortic. Sci.*, **97**: 599–603.

Grosser, J. W. and Gmitter, F. G. Jr (1990). Somatic hybridization of *Citrus* with wild relatives for germplasm enhancement and cultivar development. *HortScience*, **25**: 147–51.

Grosser, J. W., Gmitter, F. G. Jr and Chandler, J. L. (1988a). Intergeneric somatic hybrid plants from sexually incompatible woody species: *Citrus sinensis* and *Severinia disticha*. *Theor. Appl. Genet.*, **75**: 397–401.

Grosser, J. W., Gmitter, F. G. Jr and Chandler, J. L. (1988b). Intergeneric somatic hybrid plants of *Citrus sinensis* cv. Hamlin and *Poncirus trifoliata* cv. Flying Dragon. *Plant Cell Rep.*, **7**: 5–8.

Grosser, J. W., Moore, G. A. and Gmitter, F. G. Jr (1989). Interspecific somatic hybrid plants from the fusion of 'Key' lime (*Citrus aurantifolia*) with 'Valencia' Sweet orange (*Citrus sinensis*) protoplasts. *Sci. Hort.*, **39**: 23–9.

Grosser, J. W., Gmitter, F. G. Jr, Tusa N. and Chandler, J. L. (1990). Somatic hybrid plants from sexually incompatible woody species: *Citrus reticulata* and *Citropsis gilletiana*. *Plant Cell Rep.*, **8**: 656–9.

Grosser, J. W., Gmitter, F. G. Jr, Sesto, F., Deng, X. X. and Chandler, J. L. (1992). Six new somatic citrus hybrids and their potential for cultivar improvement. *J. Am. Soc. Hort. Sci.*, **117**: 169–73.

Guerra, M. dos S. (1984a). New chromosome numbers in Rutaceae. *Plant Syst. Evol.*, **146**: 13–30.

Guerra, M. dos. S. (1984b). Cytogenetics of Rutaceae. II. Nuclear DNA content. *Caryologia*, **37**: 219–26.

Hearn, C. J. (1984). Development of seedless orange and grapefruit cultivars through seed irradiation. *J. Am. Soc. Hort. Sci.*, **109**: 270–3.

Hearn, C. J. (1986). Development of seedless grapefruit cultivars through budwood irradiation. *J. Am. Soc. Hort. Sci.*, **111**: 304–6.

Hearn, C. J. (1989). Yield and fruit quality of 'Ambersweet' orange hybrid on different rootstocks. *Proc. Fla. State Hort. Soc.*, **102**: 75–8.

Hensz, R. A. (1971). 'Star Ruby' a new deep red fleshed grapefruit variety with distinct tree characteristics. *J. Rio Grande Valley Hort. Soc.*, **25**: 54–8.

Hensz, R. A. (1985). 'Rio Red' a new grapefruit with a deep-red color. *J. Rio Grande Valley Hort. Soc.*, **38**: 75–8.

Hidaka, T. (1984). Induction of plantlets from anthers of Trovita orange (*Citrus sinensis* Osbeck). *J. Japan. Soc. Hort. Sci.*, **53**: 1–5.

Hidaka, T. and Omura, M. (1989). Origin and development of embryoids from microspores in anther culture of citrus. *Japan. J. Breed.*, **39**: 169–78.

Hidaka, T. and Omura, M. (1993). Expression of GUS gene in *Citrus* transformants and the transient expression in protoplasts In *Techniques on Gene Diagnosis and Breeding in Fruit Trees*, ed. T. Hayashi, M. Omura and N. Scott, pp. 193–205. Tsukuba, Japan: FTRS.

Hidaka, T., Yamada, Y. and Shichijo, T. (1979). *In vitro* differentiation of haploid plants by anther culture in *Poncirus trifoliata* (L.) Raf. *Japan. J. Breed.*, **29**: 248–54.

Hidaka, T., Yamada, Y. and Shichijo, T. (1982). Plantlet formation by anther culture of *Citrus aurantium*. *Japan. J. Breed*, **32**: 247–52.

Hidaka, T., Omura, M., Ugaki, M., Tomiyama, M., Kato, M., Oshima, M. and Motoyoshi, F. (1990). *Agrobacterium* mediated transformation and regeneration of *Citrus* spp. from suspension cells. *Japan. J. Breed.*, **40**: 199–207.

Hirai, M., Kozaki, I. and Kajiura, I. (1986). The rate of spontaneous inbreeding of trifoliate orange and some characteristics of the inbred seedlings. *Japan. J. Breed.*, **36**: 138–46.

Hutchinson, D. J., O'Bannon, J. H., Grimm, G. R. and Bridger, G. D. (1972). Reaction of selected citrus rootstocks to foot rot and burrowing citrus nematodes, *Proc. Fla. State Hort. Soc.*, **85**: 39–43.

Ikeda, F. (1982). Repression of polyembryony by gamma-rays in polyembryonic citrus. In *Proc. Int. Soc. Citriculture, 1981*, ed. K. Matsumoto, pp. 39–44. Okitsu, Shizuoka, Japan: Okitsu Fruit Tree Research Station.

Iwamasa, M. (1966). Study on the sterility in genus *Citrus* with special reference to the seedlessness. *Bull. Hort. Res. Sta. Japan. Ser. B*, **6**: 2–77.

Iwamasa, M. (1967). Breeding of seedless citrus variety. *Japan. Agric. Res. Q.*, **2**: 19–20.

Iwamasa, M. and Oba, Y. (1975). Precocious flowering of citrus seedlings. Part I. *Agric. Bull. Saga Univ.*, **39**: 45–56. (Japanese with English summary.)

Iwamasa, M. and Oba, Y. (1980). Seedlessness due to self-compatibility in Egami-Buntan, a Japanese pummelo cultivar. *Agric. Bull. Saga Univ.*, **49**: 39–45. (Japanese with English summary.)

Iwamasa, M., Ueno, I. and Nishiura, M. (1970). Location of zygotic embryo in polyembryonic citrus seed. *Bull. Hort. Res. Sta., Japan*, Ser. B, No. 10. (English summary, 16 pp.)

Iwamasa, M., Nishiura, M., Okudai, N. and Ishiuchi, D. (1978). Characteristics due to chimeras and their stability in citrus cultivars. In *Proc. Int. Soc. Citriculture, 1977*, ed. W. Grierson, pp. 571–4. Lake Alfred, FL: ISC.

Jarrell, D. C., Roose, M. L., Traugh, S. N. and Kupper, R. S. (1992). A genetic map of citrus based on the segregation of isozymes and RFLPs in an intergeneric cross. *Theor. Appl. Genet.*, **84**: 49–56.

Jefferson, R. A., Kavanagh, T. A. and Bevan, M. V. (1987). GUS fusion glucuronidase is a sensitive and versatile fusion marker in higher plants. *EMBO J.*, **6**: 3091–7.

Juarez, J., Navarro, L. and Guardiola, J. L. (1976). Obtention des plantes nucellaires de divers cultivars de clementiniers au moyen de la culture de nucelle *in vitro*. *Fruits*, **31**: 751–62.

Kaplan, D. and O'Bannon, J. H. (1981). Evaluation and nature of citrus nematode resistance in Swingle citrumelo. *Proc. Fla. Hort. Soc.*, **94**: 33–6.

Kobayashi, S. (1987). Uniformity of plants regenerated from orange (*Citrus sinensis* Osb.) protoplasts. *Theor. Appl. Genet.*, **74**: 10–14.

Kobayashi, S. and Ohgawara, T. (1988). Production of somatic hybrid plants through protoplast fusion in citrus. *Japan Agric. Res. Q.*, **22**: 181–8.

Kobayashi, S. and Uchimiya, H. (1989). Expression and integration of a foreign gene in orange (*Citrus sinensis* Osb.) protoplasts by direct DNA transfer. *Japan. J. Genet.*, **64**: 91–7.

Kobayashi, S., Ikeda, I. and Nakatani, M. (1979). Studies on nucellar embryogenesis in Citrus II. *J. Japan. Soc. Hort. Sci.*, **48**: 179–85.

Kobayashi, S., Uchimiya, H. and Ikeda, I. (1983). Plant regeneration from 'Trovita' orange protoplasts. *Japan. J. Breed.*, **33**: 119–22.

Kobayashi, S., Fujiwara, K., Oiyama, I., Ohgawara, T. and Ishii, S. (1988a). Somatic hybridization between Navel orange and 'Murcott' tangor. In *Proc. Sixth International Citrus Congress, 1988*, ed. R. Goren and K. Mendel, pp. 135–40. Philadelphia/Rehovot: Balaban; Weikersheim, Germany: Margraf Scientific Books.

Kobayashi, S., Ohgawara, T., Ohgawara, E., Oiyama, I. and Ishii, S. (1988b). A somatic hybrid plant obtained by protoplast fusion between navel orange (*Citrus sinensis*) and satsuma mandarin. *Plant, Cell, Tissue and Organ Cult.*, **14**: 63–9.

Kochba, J. and Spiegel-Roy, P. (1973). Effects of culture media on embryoid formation from ovular callus of 'Shamouti' orange (*Citrus sinensis*). *Z. Pflanzenzuecht.*, **69**: 156–62.

Kochba, J., Spiegel-Roy, P. and Safran, H. (1972). Adventive plants from ovules and nucelli in *Citrus*. *Planta*, **106**: 237–45.

Kochba, J., Spiegel-Roy, P., Saad, S. and Neuman, H. (1978). Stimulation of embryogenesis in citrus tissue culture by galactose. *Naturwissenschaften*, **65**: 261.

Kochba, J., Ben-Hayyim, G., Spiegel-Roy, P., Neuman, H. and Saad, S. (1982a). Selection of stable salt tolerant callus cell lines and embryos in *C. sinensis* and *C. aurantium*. *Z. Pflanzenphysiol.*, **106**: 111–18.

Kochba, J., Spiegel-Roy, P., Neuman, H. and Saad, S. (1982b). Effect of carbohydrates on somatic embryogenesis in subcultured nucellar callus of Citrus cultivars. *Z. Pflanzenphysiol.*, **105**: 359–68.

Koizumi, M. and Kuhara, S. (1982). Evaluation of citrus plants for resistance to bacterial canker disease in relation to the lesion extension. *Bull. Fruit Tree Res. Sta.*, Ser. D, **4**: 73–92.

Koltunow, A. M. (1993). Isolation and construction of genes to control seed production in *Citrus*. In *Techniques on Gene Diagnosis and Breeding in Fruit Trees*, ed. T. Hayashi, M. Omura, and N. S. Scott, pp. 101–8, Tsukuba, Japan: FTRS.

Kunitake, H., Kagami, H. and Mii, M. (1991). Somatic embryogenesis and plant regeneration from protoplasts of Satsuma/mandarin (*Citrus unshiu* Marc.). *Scientia Horticulturae*, **47**: 27–33.

Lapins, K. O. (1983). Mutation Breeding. In *Methods in Fruit Breeding*, ed. M. J. Moore and J. Janick, pp. 94–9. West Lafayette, Indiana: Purdue University Press.

Lee, C. H. and Power, J. B. (1988). Intraspecific gametosomatic hybridization in *Petunia* hybrids. *Plant Cell Rep.*, **7**: 17–18.

Lee, L. S. (1988). Citrus polyploidy origin and potential for cultivar improvement. *Austr. J. Agr. Res.*, **39**: 735–47.

Ligeng, C., Keling, C., Jun, L., and Yunguan, H. (1993). Studies on the rind color heredity of Citrus. *Acta Hort. Sinica*, **29**: 221–4. (English summary.)

Ling, J. T., Nito, N. and Iwamasa, M. (1989). Plant regeneration from protoplasts of Calamondin (*Citrus madurensis* Lour.) *Sci. Hort.*, **40**: 325–33.

Litz, R. E., Moore, G. A. and Srinivasan, C. (1985). *In vitro* systems for propagation and improvement of tropical fruits and palms. *Hort. Rev.*, **7**: 157–200.

Louzada, E. S., Grosser, J. W., Gmitter, F. G. Jr., Deng, X.X., Tusa, N., Nielsen, B. and Chandler, J. L. (1992). Eight new somatic rootstocks with potential for improved disease resistance. *Hort. Sci.*, **27**: 1033–6.

Maheshwari, P. and Rangaswamy, N. S. (1958). Polyembryony and *in vitro* culture of embryos of *Citrus* and *Mangifera*. *Indian J. Hort.*, **15**: 275–82.

Matsumoto, R. and Okudai, N. (1991). Inheritance of flavanone neohesperidoside in Citrus. *J. Japan. Soc. Hort. Sci.*, **60**: 201–7.

Moore, G. A., DeWald, M. G. and Cline, K. (1989). *Agrobacterium* mediated transformation of *Citrus*. *J. Cell Biochem. Suppl. 13D*, 225 pp. (Abstract.)

Moore, G. A., Jacono, C. C., Neidigh, J. L., Lawrence, S. D. and Cline, K. (1992). Agrobacterium-mediated transformation of *Citrus* stem segments and regeneration of transgenic plants. *Plant Cell Rep.*, **11**: 238–42.

Nakatani, M., Ikeda, I. and Kobayashi, S. (1982). Studies on an effective method for getting hybrid seedlings in polyembryonic citrus 3. Artificial control of the number of embryos per seed on Minneola tangelo and sweet orange cultivars by high temperature treatment. *Bull. Fruit Tree Res. Sta. Akitsu, Ser E*, **4**: 29–40. (Japanese with English summary.)

Nati, P. (1929). Memoria sull'agrume bizzarria. Ristampa integrale del testo latino con traduzione e note del Dott A. Ragionieri. Francesco Battiato, Catania. 64 pp. (Latin reprint of 1674 Florence edition and Italian translation.)

Navarro, L. (1982). Citrus shoot tip grafting *in vitro* (STG) and its applications: a review. In *Proc. Int. Soc. Citriculture, 1981*, ed. K. Matsumoto, pp. 452–6. Okitsu, Shizuoka, Japan: Okitsu Fruit Tree Research Station.

Navarro, L., Roistacher, C. N. and Murashige, T. (1975). Improvement of shoot-tip grafting *in vitro* for virus-free citrus. *J. Am. Soc. Hort. Sci.*, **100**: 471–9.

Neilson-Jones, W. (1969). *Plant chimeras*, 2nd edn. London: Methuen, 123 pp.

Nishiura, M., and Iwamasa, M. (1970). Reversion of fruit color in nucellar seedlings from the Dobashibeni Unshu, a red color mutant of the Satsuma mandarin. *Bull. Hort. Res. Sta., Ser. B*, **10**: 1–5. (Japanese with English summary.)

Nito, N. and Iwamasa, M. (1990). *In vitro* plantlet formation from juice vesicles of callus of satsuma (*Citrus unshiu* Marc.) *Plant, Cell Tissue Org. Cult.*, **20**: 137–40.

Ohgawara, T. and Kobayashi, S. (1991). Application of protoplast fusion to *Citrus* breeding. *Food Biotechnol.*, **5**: 169–84.

Ohgawara, T., Kobayashi, S., Ohgawara, E., Uchimiya, H. and Ishii, S. (1985). Somatic hybrid plants obtained by protoplast fusion between *Citrus sinensis* and *Poncirus trifoliata*. *Theor. Appl. Genet.*, **71**: 1–4.

Ohgawara, T., Kobayashi, S., Ishii, S., Yashinaga, K. and Oiyama, I. (1989). Somatic hybridization in Citrus: navel orange (*C. sinensis* Osb.) and grapefruit (*C. paradisi* Macf.) *Theor. Appl. Genet.*, **78**: 609–12.

Ohgawara, T., Kobayashi, S., Ishii, S., Yashinaga, K. and Oiyama, I. (1991). Fertile fruit trees obtained by somatic hybridization: navel orange (*Citrus sinensis*) and Troyer citrange (*C. sinensis* × *Poncirus trifoliata*). *Theor. Appl. Genet.*, **81**: 141–3.

Oiyama, I. and Okudai, N. (1986). Production of colchicine induced autotetraploid plants through micrografting in monoembryonic citrus cultivars. *Japan. J. Breed.*, **36**: 371–6.

Ollitrault, P. and Michaux-Ferriere, N. (1994). Application of flow cytometry for Citrus genetics and breeding. In *Proc. Int. Soc. Citriculture, 1992*, ed. E. Tribulato, A. Gentile and G. Reforgiato, Vol. 1, pp. 93–198. Catania, Italy: MSC Congress.

Pirrie, A. and Power, J. B. (1986). The production of fertile, triploid somatic hybrid plants (*Nicotiana glutinosa* (n) + N. tabacum (2n)) via gametic: somatic protoplast fusion. *Theor. Appl. Genet.*, **72**: 48–52.

Rangan, T. S., Murashige, T. and Bitters, W. P. (1969). *In vitro* initiation of nucellar embryos in monoembryonic citrus. *Hort. Sci.*, **3**: 226–7.

Reem, C. L. and Furr, J. R. (1976). Salt tolerance of some *Citrus* species, relatives and hybrids tested as rootstocks. *J. Am. Soc. Hort. Sci.*, **101**: 265–7.

Roose, M. L. and Traugh, S. N. (1988). Identification and performance of citrus on nucellar and zygotic rootstocks. *J. Am. Hort. Sci.*, **113**: 100–5.

Roose, M. L., Cheng, F. S. and Federici, C. T. (1994). Origin, inheritance and effects of a dwarfing gene from the citrus rootstock *Poncirus trifoliata* 'Flying Dragon'. *HortScience*, **29**: 482 (Abstract).

Shamel, A. D. (1943). Bud variations and bud selection. In *The Citrus Industry I*, 1st edn., ed. H. J. Webber and L. D. Batchelor, pp. 915–52. Berkeley and Los Angeles: University California Press.

Shinozaki, S., Fujita, K., Hidaka, T. and Omura, M. (1992). Plantlet formation of somatic hybrids of sweet orange (*Citrus sinensis*) and its wild relative orange jessamine (*Murraya paniculata*) by electrically-induced protoplast fusion. *Japan. J. Breed.*, **42**: 287–95.

Snowball, A. M., Zeman, A. M., Tchan, Y. T., Mullins, M. G. and Goodwin, P. B. (1991). Phase change in *Citrus*: immunologically detectable differences between juvenile and mature plants. *Austr. J. Plant Physiol.*, **18**: 385–96.

Soost, R. K. (1969). The incompatibility gene system in *Citrus*. In *Proc. First Int. Citrus Symp., 1968*, ed. H. Chapman, pp. 189–90. Riverside: University of California.

Soost, R. K. (1987). Breeding, citrus genetics and nucellar embryony. In *Improving Vegetatively Propagated Crops*, ed. A. J. Abbott and R. K. Atkin, pp. 83–110. London: Academic Press.

Soost, R. K. and Cameron, J. W. (1975). Citrus. In *Advances in Fruit Breeding*, ed. J. Janick and J. N. Moore, pp. 507–40. West Lafayette, Indiana: Purdue University Press.

Soost, R. K. and Cameron, J. W. (1980). 'Oroblanco' a triploid pummelo–grapefruit hybrid. *HortScience*, **15**: 667–9.

Soost, R. K. and Cameron, J. W. (1985). 'Melogold', a triploid pummelo–grapefruit hybrid. *HortScience*, **20**: 1134–5.

Soost, R. K., Williams, T. E. and Torres, A. M. (1980). Identification of nucellar and zygotic seedlings with leaf isozymes. *HortScience*, **15**: 728–9.

Spiegel-Roy, P. (1979). On the chimeral nature of the Shamouti orange. *Euphytica*, **28**: 361–5.

Spiegel-Roy, P. (1990). *Economic and Agricultural Impact of Mutation Breeding in Fruit Trees*. Mutation Breeding Review, May 1990, Vol. 5, 26 pp. Vienna: IAEA.

Spiegel-Roy, P. and Ben-Hayyim, G. (1985). Selection and breeding for salinity tolerance *in vitro*. *Plant and Soil*, **89**: 243–52.

Spiegel-Roy, P. and Teich, A. H. (1972). Thorn as a possible genetic marker to distinguish zygotic from nucellar seedlings in citrus. *Euphytica*, **21**: 534–7.

Spiegel-Roy, P. and Vardi, A. (1984). *Citrus*. In *Handbook of Plant Cell Culture*, Vol. 3, ed. P. V. Ammirato, D. A. Evans, W. R. Sharp, and Y. Yamada, pp. 355–72. New York, London: Macmillan.

Spiegel-Roy, P. and Vardi, A. (1987). Niva and Edit – two new selections from our breeding program. *Proc. Int. Soc. Citriculture, 1984*, Vol. 1, ed. H. W. S. Montenegro and C. S. Moreira, pp. 65–7. Sao Paolo, Brazil: Inst. Econ. Agricola, Centro Estadual.

Spiegel-Roy, P. and Vardi, A. (1989). Induced mutations in Citrus. In *Proc. 6th International Congress*, pp. 773–6, Tokyo: SABRAO.

Spiegel-Roy, P., Kochba, J. and Saad, S. (1983). Selection for tolerance to 2,4-dichlorophenoxyacetic acid in ovular callus of orange (*Citrus sinensis*). *Z. Pflanzenphysiol.*, **109**: 41–8.

Spiegel-Roy, P., Vardi, A. and Elhanati, A. (1985). Seedless induced mutant in lemon (*Citrus limon*). *Mutation Breed. Newsl.*, **26**: 1.

Spiegel-Roy, P., Vardi, A., Elhanati, A., Solel, Z. and Bar-Joseph, M. (1988). Rootstock selection from a Poorman × *Poncirus trifoliata* cross. In *Proc. Sixth Intern. Citrus Congress*, ed. R. Goren, and K. Mendel, pp. 195–200. Philadelphia/Rehovot: Balaban Publishers; Weikersheim, Germany: Margraf Scientific Books.

Spiegel-Roy, P., Vardi, A. and Elhanati, A. (1990). Seedless induced mutant in highly seeded lemon (*Citrus limon*). *Mutation Breed. Newsl.*, **36**: 11.

Starrantino, A. and Russo, R. (1980). Seedlings from undeveloped ovules of ripe fruits of polyembryonic citrus cultivars. HortScience, 15: 296–7.
Strassburger, E. (1907). Ueber die individualitat der Chromosomen und die Pfropfhybriden–Frage. Jahrb. Wiss. Bot., 44: 482–555.
Sudahono, Byrne, D. H. and Rouse, R. E. (1994). Greenhouse screening of Citrus rootstocks for tolerance to bicarbonate-induced iron chlorosis. HortScience, 29: 113–16.
Swingle, W. T. and Reece, P. C. (1967). The botany of citrus and its wild relatives. In Citrus Industry, Vol. I, ed. W. Reuther, H. J. Webber and L. D. Batchelor, pp. 190–430. Berkeley: Division of Agricultural Science, University of California.
Sykes, S. R. (1985). A glasshouse screening procedure for identifying citrus hybrids which restrict chloride accumulation in shoot tissues. Austr. J. Agric. Res., 36: 779–89.
Tachikawa, T. (1971). Investigations on the breeding of citrus trees. IV. On the chromosome numbers in citrus. Bull. Shizuoka Prefect. Citrus Exp. Stn., 9: 11–25. (Japanese with English Summary.)
Takayanagi, R., Hidaka, T. and Omura, M. (1992). Regeneration of intergeneric somatic hybrids by electrical fusion between Citrus and its wild relatives Mexican lime (Citrus aurantifolia and Java Feroniella (Feroniella lucida) or Tabog (Swinglea glutinosa). Japan. Soc. Hort. Sci., 60: 799–804.
Tanaka, T. (1927). Bizzarria: a clear case of periclinal chimera. J. Genet., 18: 77–85.
Tatum, J. H., Hearn, C. J. and Berry, R. E. (1978). Characterization of citrus cultivars by chemical differentiation. J. Am. Soc. Hort. Sci., 103: 492–6.
Tilney-Bassett, R. A. E. (1986). Plant Chimeras. London: Edward Arnold, 199 pp.
Torres, A. M., Soost, R. K. and Mau-Lastovicka, T. (1982). Citrus isozymes. Genetics and distinguishing nucellar zygotic seedlings. J. Hered., 73: 335–9.
Tusa, N., Grosser, J. W. and Gmitter, F. G. (1990). Plant regeneration of 'Valencia' sweet orange, 'Femminello' lemon and the interspecific somatic hybrid following protoplast fusion. J. Am. Soc. Hort. Sci., 115: 1043–6.
Ueno, I. (1978). Studies of cross-incompatibility in Citrus tachibana Tanaka. I. Fruit set of Tachibana after cross pollination with eight citrus varieties. Bull. Fruit Tree Res. Stn., Ser B. (Akitsu, Japan), 5: 1–7.
Vardi, A. and Galun, E. (1988). Recent advances in protoplast culture of horticultural crops: Citrus. Sci. Hort., 37: 217–30.
Vardi, A. and Spiegel-Roy, P. (1978). Taxonomy, breeding and genetics. In Proc. Int. Soc. Citriculture 1977, ed. W. Grierson, pp. 51–7. Lake Alfred, FL: ISC.
Vardi, A. and Spiegel-Roy, P. (1982). Gene control in meiosis of Citrus reticulata. In Proc. Int. Soc. Citriculture 1981, ed. K. Matsumoto, pp. 26–7. Okitsu, Shizuoka, Japan: Okitsu Fruit Tree Research Station.
Vardi, A. and Spiegel-Roy, P. (1988). A new approach to selection for seedlessness. In Proceedings of the Sixth International Citrus Congress 1988, ed. R. Goren and K. Mendel, pp. 131–4. Philadelphia/Rehovot: Balaban Publishers; Weikersheim, Germany: Margraf Scientific Books.
Vardi, A., Spiegel-Roy, P. and Galun, E. (1975). Citrus cell culture:

isolation of protoplasts, planting densities, effect of mutagens and regenerations of embryos. *Plant Sci. Lett.*, **4**: 231–6.

Vardi, A., Spiegel-Roy, P. and Galun, E. (1982). Plant regeneration from citrus protoplasts: variability in methodological requirements among cultivars and species. *Theor. Appl. Genet.* **62**: 171–6.

Vardi, A., Hutchison, D. J. and Galun, E. (1986). A protoplast to tree system in *Microcitrus* based on protoplasts derived from a sustained embryogenic callus. *Plant Cell Rep.*, **5**: 412–14.

Vardi, A., Breiman, A. and Galun, E. (1987). *Citrus* cybrids: production by donor–recipient protoplast fusion and verification by mitochondrial-DNA restriction profiles. *Theor. Appl. Genet.*, **75**: 51–8.

Vardi, A., Spiegel-Roy, P., Ben-Hayyim, G., Neumann, H., and Shalhevet, J. (1988). Responses of Shamouti orange and Minneola tangelo on six rootstocks to salt stress. In *Proc. Sixth Int. Citrus Congr.*, *1988*, ed. R. Goren and K. Mendel, pp. 75–82. Philadelphia/Rehovot: Balaban Publishers, Weikersheim, Germany: Margraf Scientific Books.

Vardi, A., Arzee-Gonen, P., Frydman-Shani, A., Bleichman, S. and Galun, E. (1989). Protoplast fusion mediated transfer of orgenelles from *Microcitrus* and *Citrus* and regeneration of novel alloplasmic trees. *Theor. Appl. Genet.*, **78**: 741–7.

Vardi, A., Bleichman, S. and Aviv, D. (1990). Genetic transformation of *Citrus* protoplasts and regeneration of transgenic plants. *Plant Sci.*, **69**: 199–206.

Wakana, S., Iwamasa, M. and Uemoto, S. (1982). Seed development in relation to ploidy of zygotic embryo and endosperm in polyembryonic citrus. In *Proc. Int. Soc. Citriculture 1981*, ed. K. Matsumoto, pp. 35–9. Okitsu, Shizuoka, Japan: Okitsu Fruit Tree Research Station.

Wang, D. and Chang, C. J. (1978). Triploid citrus plantlet from endosperm culture. *Sci. Sinica*, **21**: 822–7. (In Chinese.)

Weinbaum, S., Cohen, E., and Spiegel-Roy, P. (1982). Rapid screening of 'Satsuma' mandarin progeny to distinguish nucellar and zygotic seedlings. *HortScience*, **17**: 239–40.

Wendel, J. F. and Weeden, N. F. (1989). Visualisation and interpretation of plant isoenzymes. In *Isoenzymes in Plant Biology*, ed. D. E. Soltis and R. J. Soltis, pp. 5–45. Portland, OR: Dioscorides Press.

Winkler, H. (1908). Ueber Pfropfbastarde und pflanzliche Chimaeren. *Ber. Deut. Bot. Gesell*, **25**: 568–76.

Yadav, I. S., Jalikop, S. H. and Singh, H. P. (1980). Recognition of short juvenility in *Poncirus*. *Curr. Sci.* (Bangalore), **49**: 512–13.

Yelenosky, G. (1985). Cold hardiness in Citrus. In *Horticultural Reviews*, ed. J. Janick, Vol. 7, pp. 201–38. Fairfield, CT: Avi Publishing Co.

Yelenosky, G., Hutchinson, D. and Barrett, H. (1993). Freezing resistance of progeny from open pollinated pummelox trifoliate orange citrus hybrids. *HortScience*, **28**: 1120–1.

Zhou, J. *et al.* (1986). Induction of seedless mutation by irradiating citrus seeds with ^{60}Co gamma rays. *China Citrus.* **2**: 1–4. (In Chinese.)

Index

abscisic acid (ABA), 105, 150
abscission, 56, 97–8
  autumnal, 47
  flower, 84
  fruit, 97–8
  fruitlet, 97, 114, 115
  leaf, 56, 98
  pedicel, 104
  shoot-tip, 50, 52
  style, 84
abscission zone, 56, 98
acaricides, selective, 162, 167
acid, 101
  citric, 101, 110
  malic, 110
  malonic, 110
acidity, 101, 110
acid lime, 43, 50
  see also Citrus aurantifolia; C. latifolia
Aegle, 20, 37
  marmelos (Bael fruit), 38
Aeglopsis, 20, 37
Afraegle, 20, 37
Agrobacterium tumefaciens, 210
agrochemicals, 104, 150
air layering (marcottage), 5, 127
albedo, 89, 90
Alemow, 129
  see also Citrus macrophylla
Aleurocanthus woglumi, 165
alternate bearing, 97, 113, 117
American grasshopper, *see Schistocerca americana*
ammonia, 73
androgenesis, 203–4
anthers, 80–1, 84–5
anthesis, 70, 76–7, 81, 91, 93–4
anthocyanins, 16, 111, 197, 199
antipodal cells, 84
Aonidiella aurantii, 161, 163, 165
Apate monachus, 163

Apate spiraecola, 155
Aphytis holoxanthus, 166
Aphytis, spp. 167
apical dominance, 76
apical meristem, 53, 76, 91
apomictic embryo, 188–90
apomixis, 34, 87, 127, 188–90
apoplastic transport, 92
archesporium, 81, 83
arrowhead scale, *see Unaspis yanonensis*
ascorbic acid, 107–8, 110
assimilates, *see* photosynthate
assimilation, *see* photosynthesis
asynapsis, 83, 197–8
Atalantia, 20, 22–4
  A. ceylanica, 24
  A. hainanensis, 24
  A. guillaumini, 24
attractants, 164–5
Aurantioideae, 19–22, 37
autocidal control, *see* pest management, autocidal
auxin(s), 63, 75, 97, 105, 150–1
  synthetic, 97, 105, 151
axillary bud, 52, 55

Bactrocera dorsalis, 164
bait sprays, 162–3, 166
Baladi mandarin, 41
Balsamocitrinae, 20–1
Balsamocitrus, 20
bayberry whitefly, *see Parabemisia myricae*
Beledi orange, 188
bergamot, 13, 84
  see also Cirus bergamia
biological control, *see* pest management, biological
biological zero, *see* temperature, biological zero
biotechnology, 36–7, 198, 204–10
black giant bostrychid, *see Apate monachus*

Printed in the United States
By Bookmasters